A MATHEMATICAL VIEW OF INTERIOR-POINT METHODS IN CONVEX OPTIMIZATION

MPS/SIAM Series on Optimization

This series is published jointly by the Mathematical Programming Society and the Society for Industrial and Applied Mathematics. It includes research monographs, textbooks at all levels, books on applications, and tutorials. Besides being of high scientific quality, books in the series must advance the understanding and practice of optimization and be written clearly, in a manner appropriate to their level.

Editor-in-Chief
John E. Dennis, Jr., Rice University

Continuous Optimization Editor
Stephen J. Wright, Argonne National Laboratory

Discrete Optimization Editor
David B. Shmoys, Cornell University

Editorial Board
Daniel Bienstock, Columbia University
John R. Birge, Northwestern University
Andrew V. Goldberg, InterTrust Technologies Corporation
Matthias Heinkenschloss, Rice University
David S. Johnson, AT&T Labs - Research
Gil Kalai, Hebrew University
Ravi Kannan, Yale University
C. T. Kelley, North Carolina State University
Jan Karel Lenstra, Technische Universiteit Eindhoven
Adrian S. Lewis, University of Waterloo
Daniel Ralph, The Judge Institute of Management Studies
James Renegar, Cornell University
Alexander Schrijver, CWI, The Netherlands
David P. Williamson, IBM T.J. Watson Research Center
Jochem Zowe, University of Erlangen-Nuremberg, Germany

Series Volumes
Renegar, James, *A Mathematical View of Interior-Point Methods in Convex Optimization*
Ben-Tal, Aharon and Nemirovski, Arkadi, *Lectures on Modern Convex Optimization: Analysis, Algorithms, and Engineering Applications*
Conn, Andrew R., Gould, Nicholas I. M., and Toint, Phillippe L., *Trust-Region Methods*

A MATHEMATICAL VIEW OF INTERIOR-POINT METHODS IN CONVEX OPTIMIZATION

James Renegar
Cornell University
Ithaca, New York

Society for Industrial and Applied Mathematics
Philadelphia

MPS
Mathematical Programming Society
Philadelphia

Copyright ©2001 by the Society for Industrial and Applied Mathematics.

10 9 8 7 6 5 4 3 2 1

All rights reserved. Printed in the United States of America. No part of this book may be reproduced, stored, or transmitted in any manner without the written permission of the publisher. For information, write to the Society for Industrial and Applied Mathematics, 3600 University City Science Center, Philadelphia, PA 19104-2688.

Library of Congress Cataloging-in-Publication Data
Renegar, James, 1955-
  A mathematical view of interior-point methods in convex optimization / James Renegar.
     p. cm. – (MPS-SIAM series on optimization)
  Includes bibliographical references and index.
  ISBN 0-89871-502-4 (pbk.)
     1. Interior-point methods. 2. Mathematical optimization. 3. Convex programming. I. Title. II. Series.

QA402.5 .R46 2001
519.3–dc21
                                                                                2001042999

This research was supported by NSF grant CCR-9403580.

 is a registered trademark.

Contents

Preface		vii
1	**Preliminaries**	**1**
	1.1 Linear Algebra	2
	1.2 Gradients	5
	1.3 Hessians	9
	1.4 Convexity	11
	1.5 Fundamental Theorems of Calculus	14
	1.6 Newton's Method	18
2	**Basic Interior-Point Method Theory**	**21**
	2.1 Intrinsic Inner Products	21
	2.2 Self-Concordant Functionals	23
	2.2.1 Introduction	23
	2.2.2 Self-Concordancy and Newton's Method	27
	2.2.3 Other Properties	31
	2.3 Barrier Functionals	35
	2.3.1 Introduction	35
	2.3.2 Analytic Centers	38
	2.3.3 Optimal Barrier Functionals	39
	2.3.4 Other Properties	40
	2.3.5 Logarithmic Homogeneity	41
	2.4 Primal Algorithms	43
	2.4.1 Introduction	43
	2.4.2 The Barrier Method	45
	2.4.3 The Long-Step Barrier Method	49
	2.4.4 A Predictor-Corrector Method	52
	2.5 Matters of Definition	54
3	**Conic Programming and Duality**	**65**
	3.1 Conic Programming	65
	3.2 Classical Duality Theory	68
	3.3 The Conjugate Functional	75
	3.4 Duality of the Central Paths	81

3.5		Self-Scaled (or Symmetric) Cones	83
	3.5.1	Introduction	83
	3.5.2	An Important Remark on Notation	85
	3.5.3	Scaling Points	86
	3.5.4	Gradients and Norms	90
	3.5.5	A Useful Theorem	95
3.6		The Nesterov–Todd Directions	97
3.7		Primal-Dual Path-Following Methods	102
	3.7.1	Measures of Proximity	102
	3.7.2	An Algorithm	104
	3.7.3	Another Algorithm	106
3.8		A Primal-Dual Potential-Reduction Method	108
	3.8.1	The Potential Function	108
	3.8.2	The Algorithm	110
	3.8.3	The Analysis	111

Bibliography   115

Index   117

Preface

This book aims at developing a thorough understanding of the most general theory for interior-point methods, a class of algorithms for convex optimization problems. The study of these algorithms has dominated the continuous optimization literature for nearly 15 years, beginning with the paper by Karmarkar [10]. In that time, the theory has matured tremendously, perhaps most notably due to the path-breaking, broadly encompassing, and enormously influential work of Nesterov and Nemirovskii [15].

Much of the literature on the general theory of interior-point methods is difficult to understand, even for specialists. My hope is that this book will make the most general theory accessible to a wide audience—especially Ph.D. students, the next generation of optimizers. The book might be used for graduate courses in optimization, both in mathematics and engineering departments. Moreover, it is self-contained and thus appropriate for individual study by students as well as by seasoned researchers who wish to better assimilate the most general interior-point method theory.

This book grew out of my lecture notes for the Ph.D. course on interior-point methods at Cornell University. The impetus for writing the book came from an invitation to give the annual lecture series at the Center for Operations Research and Econometrics (CORE) at Louvain-la-Neuve, Belgium, in fall, 1996. The book is being published by CORE as well as in the SIAM-MPS series.*

The writing of this book has been a particularly satisfying experience. It has brought into sharp focus the beauty and coherence of the interior-point method theory as developed and influenced through the efforts of many researchers. I hope those researchers will not be offended by my choice to cite few references (and I hope they largely agree that the ones I have cited are the ones that should be cited, given the material covered herein). The reader can look to recent articles and books for extensive bibliographies (cf. [21], [22], [24], [25]); a great place to stay up to date is Interior-Point Methods Online (http://www-unix.mcs.anl.gov/otc/InteriorPoint/).

In citing few references, my intent is not to give the impression of research originality. Indeed, most of the results in this monograph are undoubtedly embedded somewhere in [15], [16], and [17] (which in turn were influenced by other works). My only claim to originality is in presenting a simplifying perspective.

Ithaca, January, 2001

---

*This SIAM-MPS version is slightly more polished than the CORE version, having benefited from reviewers as well as from students who combed the manuscript while taking the Ph.D. course on interior-point methods. I wish to thank Osman Güler, who carefully read the manuscript in his role as a reviewer for the SIAM-MPS series. His suggestions improved the presentation. I am very grateful to Francis Su and Trevor Park, both of whom suggested many changes that benefit the reader.

Chapter 1
Preliminaries

This chapter provides a review of material generally pertinent to continuous optimization theory, although it is written so as to be readily applicable in developing interior-point method (ipm) theory. The primary difference between our exposition and more customary approaches is that we do not rely on coordinate systems. For example, it is customary to define the gradient of a functional $f : \mathbb{R}^n \to \mathbb{R}$ as the vector-valued function $g : \mathbb{R}^n \to \mathbb{R}^n$ whose jth coordinate is $\partial f / \partial x_j$. Instead, we consider the gradient as determined by an underlying inner product $\langle \ , \ \rangle$. For us, the gradient is the function g satisfying

$$\lim_{\|\Delta x\| \to 0} \frac{f(x + \Delta x) - f(x) - \langle g(x), \Delta x \rangle}{\|\Delta x\|} = 0,$$

where $\|\Delta x\| := \langle \Delta x, \Delta x \rangle^{1/2}$. In general, the function whose jth coordinate is $\partial f / \partial x_j$ is the gradient only if $\langle \ , \ \rangle$ is the Euclidean inner product.

The natural geometry varies from point to point in the domains of optimization problems that can be solved by ipm's. As the algorithms progress from one point to the next, one changes the inner product—and hence the geometry—to visualize the headway achieved by the algorithms. The relevant inner products may bear no relation to an initially imposed coordinate system. Consequently, in aiming for the most transparent and least cumbersome proofs, one should dispense with coordinate systems.

We begin with a review of linear algebra by recalling, for example, the notion of a self-adjoint linear operator. We then define gradients and Hessians, emphasizing how they change when the underlying inner product is changed. Next is a brief review of basic results for convex functionals, followed by results akin to the fundamental theorem of calculus. Although these "calculus results" are elementary and dry, they are essential in achieving lift-off for the ipm theory. Finally, we recall Newton's method for continuous optimization, proving a standard theorem which later plays a central motivational role.

Seasoned researchers will likely find parts of this chapter to be overly detailed, especially the proofs for results which are essentially only multivariate calculus. The researchers might wonder why standard calculus results are not employed for the sake of brevity. This chapter is aimed at new Ph.D. students rather than seasoned researchers. The development is meant to teach students to think coordinate-free, quite contrary to how multivariate calculus is customarily taught; the development is in the spirit of functional analysis. By patiently learning to think coordinate-free in Chapter 1, students are in a strong position to quickly grasp the essentials of the ipm theory in Chapter 2. Beginning with Chapter 2, proofs are more concise and written at a level appropriate to research.

1.1 Linear Algebra

We let $\langle\,,\,\rangle$ denote an arbitrary inner product on $\mathbb{R}^n$. In later sections, $\langle\,,\,\rangle$ will act as a reference inner product, an inner product from which other inner products are constructed. For the ipm theory, it happens that the reference inner product $\langle\,,\,\rangle$ is irrelevant; the inner products essential to the theory are independent of the reference inner product. To a large extent, the reference inner product will serve only to fix notation.

Although the particular reference inner product will prove to be irrelevant for ipm theory, for many optimization problems to be solved by ipm's there are natural reference inner products. For example, in linear programming (LP) where vectors x are expressed coordinatewise and "$x \geq 0$" means each coordinate is nonnegative, the natural inner product is the Euclidean inner product, which we refer to as the *dot product*, writing $x_1 \cdot x_2$. Similarly, in semidefinite programming (SDP) where the relevant vector space is $\mathbb{S}^{n \times n}$—the space of symmetric $n \times n$ real matrices X—and "$X \succeq 0$" means X is positive semidefinite (i.e., has no negative eigenvalues), the natural inner product is the *trace product*,

$$X_1 \circ X_2 := \text{trace}(X_1 X_2).$$

Thus, $X_1 \circ X_2$ equals the sum of the eigenvalues of the matrix $X_1 X_2$.[1]

Throughout the general development, we use $\mathbb{R}^n$ to denote an arbitrary finite-dimensional real vector space with an inner product, be it $\mathbb{S}^{n \times n}$ or whatever. (The general development can be done in the setting of arbitrary real Hilbert spaces, even infinite-dimensional ones. We restrict ourselves to $\mathbb{R}^n$ to reach a wide audience.)

The inner product $\langle\,,\,\rangle$ induces a norm on $\mathbb{R}^n$,

$$\|x\| := \langle x, x \rangle^{1/2}.$$

Perhaps the most useful relation between the inner product and the norm is the *Cauchy–Schwarz inequality*,

$$|\langle x_1, x_2 \rangle| \leq \|x_1\|\,\|x_2\|$$

with equality iff x_1 and x_2 are colinear.

Whereas the dot product gives rise to the Euclidean norm, the norm arising from the trace product is known as the *Frobenius norm* (or the *Hilbert–Schmidt norm*). The trace product, and hence the Frobenius norm, can be extended to the vector space of all real $n \times n$ matrices by defining $M_1 \circ M_2 := \text{trace}(M_1^T M_2)$ and $\|M\| := (\sum m_{ij}^2)^{1/2}$ where m_{ij} are the coefficients of M.

---

[1] If one catenates the rows of $n \times n$ symmetric matrices to form vectors in $\mathbb{R}^{n^2}$, the trace product of two matrices is identical to the dot product of their resulting vectors. This is not to say that $\mathbb{S}^{n \times n}$ endowed with the trace product is no different than $\mathbb{R}^{n^2}$ endowed with the dot product, since the vectors obtained by catenating rows of symmetric matrices form a proper and highly structured subspace of $\mathbb{R}^{n^2}$. Rather than thinking of the dot product when considering the trace product, it is best to think of eigenvalues: $X_1 \circ X_2$ is the sum of the eigenvalues of the matrix $X_1 X_2$.

1.1. Linear Algebra

We remark that the Frobenius norm is submultiplicative, meaning $\|XS\| \leq \|X\|\|S\|$, as is seen from the relations

$$\|XS\|^2 = \sum_i \sum_j \left(\sum_k x_{ik} s_{kj}\right)^2$$
$$\leq \sum_i \sum_j \left(\sum_k x_{ik}^2\right)\left(\sum_k s_{kj}^2\right)$$
$$= \|X\|^2 \|S\|^2.$$

Recall that vectors $x_1, x_2 \in \mathbb{R}^n$ are said to be *orthogonal* if $\langle x_1, x_2 \rangle = 0$. Recall that a basis $v_1, \ldots, v_n$ for $\mathbb{R}^n$ is said to be an *orthonormal* if

$$\langle v_i, v_j \rangle = \delta_{ij} \text{ for all } i, j,$$

where δ_{ij} is the Kronecker delta. A linear operator (i.e., a linear transformation) $Q : \mathbb{R}^n \to \mathbb{R}^n$ is said to be *orthogonal* if

$$\langle Qx_1, Qx_2 \rangle = \langle x_1, x_2 \rangle \text{ for all } x_1, x_2 \in \mathbb{R}^n.$$

If given an inner product $\langle\ ,\ \rangle$ on $\mathbb{R}^n$, one adopts a coordinate system obtained by expressing vectors as linear combinations of an orthonormal basis; the inner product $\langle\ ,\ \rangle$ is the dot product for that coordinate system. Consequently, one can consider the results we review below as following from the special case of the dot product, but one should keep in mind that thinking in terms of coordinates is best avoided for understanding the ipm theory.

If both $\mathbb{R}^n$ and $\mathbb{R}^m$ are endowed with inner products and $A : \mathbb{R}^n \to \mathbb{R}^m$ is a linear operator, there exists a unique linear operator $A^* : \mathbb{R}^m \to \mathbb{R}^n$ satisfying

$$\langle Ax, y \rangle = \langle x, A^*y \rangle \text{ for all } x \in \mathbb{R}^n, y \in \mathbb{R}^m.$$

The operator A^* is the *adjoint* of A. The range space of A^* is orthogonal to the nullspace of A.

If A is surjective, then A^* is injective and the linear operator $A^*(AA^*)^{-1}A$ projects $\mathbb{R}^n$ orthogonally onto the range space of A^*; that is, the image of x is the point in the range space closest to x. Likewise, $I - A^*(AA^*)^{-1}A$ projects $\mathbb{R}^n$ orthogonally onto the nullspace of A.

If both $\mathbb{R}^n$ and $\mathbb{R}^m$ are endowed with the dot product and if A and A^* are written as matrices, then $A^* = A^T$, the transpose of A.

It is a simple but important exercise for SDP to show that if $S_1, \ldots, S_m \in \mathbb{S}^{n \times n}$ and $A : \mathbb{S}^{n \times n} \to \mathbb{R}^m$ is the linear operator defined by

$$X \mapsto (X \circ S_1, \ldots, X \circ S_m),$$

then

$$A^*y = \sum_i y_i S_i,$$

assuming $\mathbb{S}^{n \times n}$ is endowed with the trace product and $\mathbb{R}^m$ is endowed with the dot product.

Continuing to assume $\mathbb{R}^n$ and $\mathbb{R}^m$ are endowed with inner products, and hence norms, one obtains an *induced operator norm* on the vector space consisting of linear operators $A : \mathbb{R}^n \to \mathbb{R}^m$:
$$\|A\| := \max\{\|Ax\| : \|x\| \le 1\}.$$

Each linear operator $A : \mathbb{R}^n \to \mathbb{R}^m$ has a *singular-value decomposition*. Precisely, there exist orthonormal bases $u_1, \ldots, u_n$ and $w_1, \ldots, w_m$, as well as real numbers $0 < \gamma_1 \le \cdots \le \gamma_r$ where r is the rank of A, such that for all x,
$$Ax = \sum_{i=1}^{r} \gamma_i \langle u_i, x \rangle w_i.$$

(Equivalently, $Au_i = \gamma_i w_i$ for $i = 1, \ldots, r$ and $Au_i = 0$ for $i > r$.) The numbers γ_i are the *singular values* of A; if $r < n$, then the number 0 is also considered to be a singular value of A. It is easily seen that $\|A\| = \gamma_r$. Moreover,
$$A^* y = \sum_{i=1}^{r} \gamma_i \langle w_i, y \rangle u_i,$$
so that the values γ_i (and possibly 0) are also the singular values of A^*. It follows that $\|A^*\| = \|A\|$.

If $\mathbb{R}^n$ and $\mathbb{R}^m$ are endowed with the dot product, the singular-value decomposition corresponds to the fact that if A is an $m \times n$ matrix, there exist orthogonal matrices Q_m and Q_n such that $Q_m A Q_n = \Gamma$ where Γ is an $m \times n$ matrix with zeros everywhere except possibly for positive numbers on the main diagonal.

It is not difficult to prove that a linear operator $Q : \mathbb{R}^n \to \mathbb{R}^n$ is orthogonal iff $Q^* = Q^{-1}$. For orthogonal operators, $\|Q\| = 1$.

A linear operator $S : \mathbb{R}^n \to \mathbb{R}^n$ is said to be *self-adjoint* if $S = S^*$.

If $\langle\,,\,\rangle$ is the dot product and S is written as a matrix, then S being self-adjoint is equivalent to S being symmetric.

It is instructive to show that for $S \in \mathbb{S}^{n \times n}$, the linear operator $A : \mathbb{S}^{n \times n} \to \mathbb{S}^{n \times n}$ defined by
$$X \mapsto SXS$$
is self-adjoint. Such operators are important in the ipm theory for SDP.

A linear operator $S : \mathbb{R}^n \to \mathbb{R}^n$ is said to be *positive semidefinite* (psd) if S is self-adjoint and
$$\langle x, Sx \rangle \ge 0 \quad \text{for all } x \in \mathbb{R}^n.$$
If, further, S satisfies
$$\langle x, Sx \rangle > 0 \quad \text{for all } x \ne 0,$$
then S is said to be *positive definite* (pd). Keep in mind that self-adjointness is a part of our definitions of psd and pd operators, contrary to the definitions found in much of the optimization literature.

Each self-adjoint linear operator S has a *spectral decomposition*. Precisely, for each self-adjoint linear operator S there exists an orthonormal basis $v_1, \ldots, v_n$ and real numbers $\lambda_1 \le \cdots \le \lambda_n$ such that for all x,
$$Sx = \sum_i \lambda_i \langle v_i, x \rangle v_i.$$

It is easily seen that v_i is an eigenvector for S with eigenvalue λ_i.

If $\langle\,,\,\rangle$ is the dot product and S is a symmetric matrix, then the spectral decomposition corresponds to the fact that S can be diagonalized using an orthogonal matrix, i.e., $Q^T S Q = \Lambda$.

The following relations are easily established:

- $\|S\| = \max_i |\lambda_i| = \max\{|\langle x, Sx\rangle| : \|x\| = 1\}$;
- S is psd iff $\lambda_i \geq 0$ for all i;
- S is pd iff $\lambda_i > 0$ for all i;
- If S^{-1} exists, then it, too, is self-adjoint and has eigenvalues $1/\lambda_i$. (In particular, $\|S^{-1}\| = 1/\min_i |\lambda_i|$.)

The spectral decomposition for a psd operator S allows one to easily prove the existence of a psd operator $S^{1/2}$ satisfying $S = (S^{1/2})^2$; simply replace λ_i by $\sqrt{\lambda_i}$ in the decomposition. In turn, the uniqueness of $S^{1/2}$ can readily be proven by relying on the fact that if T is a psd operator satisfying $T^2 = S$, then the eigenvectors for T are eigenvectors for S. The operator $S^{1/2}$ is the *square root* of S.

Here is a crucial observation: If S is pd, then S defines a new inner product, namely,

$$\langle x_1, x_2\rangle_S := \langle x_1, Sx_2\rangle.$$

Every inner product on $\mathbb{R}^n$ arises in this way; that is, regardless of the initial inner product $\langle\,,\,\rangle$, for every other inner product there exists S which is pd with respect to (w.r.t.) $\langle\,,\,\rangle$ and for which $\langle\,,\,\rangle_S$ is precisely the other inner product. (We do not rely on this fact.)

Let $\|\ \|_S$ denote the norm induced by $\langle\,,\,\rangle_S$.

Assume A^* is the adjoint of $A : \mathbb{R}^n \to \mathbb{R}^m$. Assuming S and T are pd w.r.t. the respective inner products, if the inner product on $\mathbb{R}^n$ is replaced by $\langle\,,\,\rangle_S$ and that on $\mathbb{R}^m$ is replaced by $\langle\,,\,\rangle_T$, then the adjoint of A becomes $S^{-1}A^*T$, as is easily shown. Moreover, letting $\|A\|_{S,T}$ denote the resulting operator norm, it is easily proven that

$$\|A\|_{S,T} = \|T^{1/2} A S^{-1/2}\| \quad \text{and} \quad \|A\| = \|T^{-1/2} A S^{1/2}\|_{S,T}.$$

Similarly, since for $c \in \mathbb{R}^n$, the linear function given by $x \mapsto \langle c, x\rangle$ is identical to the function $x \mapsto \langle S^{-1}c, x\rangle_S$, if c is the objective vector of an optimization problem written in terms of $\langle\,,\,\rangle$—that is, if the objective is "$\min\langle c, x\rangle$"—then when written in terms of $\langle\,,\,\rangle_S$, the objective vector is $S^{-1}c$. The notation used in this paragraph illustrates our earlier assertion that the reference inner product will be useful in fixing notation as the inner products change.

It is instructive to consider the shape of the unit ball w.r.t. $\|\ \|_S$ viewed in terms of the geometry of the reference inner product. The spectral decomposition of S easily implies the unit ball to be an ellipsoid with axes in the directions of the orthonormal basis vectors $v_1, \ldots, v_n$, the length of the axis in the direction of v_i being $2\sqrt{1/\lambda_i}$.

1.2 Gradients

Recall that a *functional* is a function whose range lies in $\mathbb{R}$. We use D_f to denote the domain of a functional f. It will always be assumed that D_f is an open subset of $\mathbb{R}^n$ in the norm topology (recalling that all norms on $\mathbb{R}^n$ induce the same topology).

Let $\langle\,,\,\rangle$ denote an arbitrary inner product on $\mathbb{R}^n$ and let $\|\ \|$ denote the norm induced by $\langle\,,\,\rangle$.

The functional f is said to be (*Fréchet*) *differentiable* at $x \in D_f$ if there exists a vector $g(x)$ satisfying
$$\lim_{\|\Delta x\| \to 0} \frac{f(x + \Delta x) - f(x) - \langle g(x), \Delta x \rangle}{\|\Delta x\|} = 0.$$
The vector $g(x)$ is the *gradient* of f at x w.r.t. $\langle\,,\,\rangle$.

If one chooses the inner product $\langle\,,\,\rangle$ on $\mathbb{R}^n$ to be the dot product and expresses $g(x)$ coordinatewise, the jth coordinate is $\partial f/\partial x_j$. This is seen by restricting Δx to be scalar multiples of e_j, the jth unit vector.

For an arbitrary inner product $\langle\,,\,\rangle$, the gradient has the same geometrical interpretation that is taught in Calculus for the dot product gradient. Roughly speaking, the gradient $g(x)$ points in the direction for which the functional output increases the fastest per unit distance traveled, and the magnitude $\|g(x)\|$ equals the amount the functional will change per unit distance traveled in that direction. (Since what constitutes "unit distance" changes as the inner product changes, it is not mysterious that the gradient changes as the inner product changes.) We give rigor to this geometrical interpretation in §1.4.

The *first-order approximation of f at x* is the linear functional
$$y \mapsto f(x) + \langle g(x), y - x \rangle.$$

If f is differentiable at each $x \in D_f$, then f is said to be *differentiable*. Henceforth, assume f is differentiable.

If the function $x \mapsto g(x)$ is continuous at each $x \in D_f$, then f is said to be *continuously differentiable*; one writes $f \in \mathcal{C}^1$.

To illustrate the definition of gradient, consider the functional
$$f(X) := -\ln \det(X)$$
with domain $\mathbb{S}^{n \times n}_{++}$, the set of all pd matrices in $\mathbb{S}^{n \times n}$. This functional plays an especially important role in SDP. We claim that w.r.t. the trace product,
$$g(X) = -X^{-1}.$$

Indeed, let $\Delta X \in \mathbb{S}^{n \times n}$ and denote the eigenvalues of $X^{-1}(\Delta X)$ by $\gamma_1, \ldots, \gamma_n$. Since $\gamma_1, \ldots, \gamma_n$ are also the eigenvalues of the symmetric matrix $X^{-1/2}(\Delta X)X^{-1/2}$, the eigenvalues are real numbers. Note that $X^{-1} \circ \Delta X = \sum_i \gamma_i$ and
$$\begin{aligned}f(X + \Delta X) - f(X) &= -\ln \det(X + \Delta X) + \ln \det(X) \\ &= -\ln \det(I + X^{-1}(\Delta X)) \\ &= -\sum_i \ln(1 + \gamma_i).\end{aligned}$$

The submultiplicativity of the Frobenius norm gives
$$\|X^{-1}(\Delta X)\|/\|X^{-1}\| \le \|\Delta X\| \le \|X^{-1}(\Delta X)\|\,\|X\|. \tag{1.1}$$

1.2. Gradients

Since $\|X^{-1}(\Delta X)\| = \|\gamma\|$ where $\|\gamma\|$ is the Euclidean norm of the vector $\gamma := (\gamma_1, \ldots, \gamma_n)$, the inequalities in (1.1) show the statement "$\|\Delta X\| \to 0$" is equivalent to "$\|\gamma\| \to 0$."

Finally, to establish the claim $g(X) = -X^{-1}$, observe that

$$\limsup_{\|\Delta X\| \to 0} \frac{|f(X + \Delta X) - f(X) - \langle -X^{-1}, \Delta X \rangle|}{\|\Delta X\|}$$

$$= \limsup_{\|\Delta X\| \to 0} \frac{|-\ln \det(X + \Delta X) + \ln \det(X) + X^{-1} \circ \Delta X|}{\|\Delta X\|}$$

$$\leq \limsup_{\|\gamma\| \to 0} \frac{\sum_i |\ln(1 + \gamma_i) - \gamma_i|}{\|\gamma\|/\|X^{-1}\|}$$

$$\leq \|X^{-1}\| \sum_i \lim_{\gamma_i \to 0} \frac{|\ln(1 + \gamma_i) - \gamma_i|}{|\gamma_i|}$$

$$= 0.$$

Our definition of what it means for a functional f to be differentiable is phrased in terms of the inner product $\langle\,,\,\rangle$. However, relying on the equivalence of all norm topologies on $\mathbb{R}^n$, it is readily proven that the property of being differentiable—and being continuously differentiable—is independent of the inner product. The gradient depends on the inner product, but differentiability does not.

The following theorem shows how the gradient changes as the inner product changes.

Theorem 1.2.1. *If S is pd and f is differentiable at x, then the gradient of f at x w.r.t. $\langle\,,\,\rangle_S$ is $S^{-1}g(x)$.*

Proof. Letting λ_1 denote the least eigenvalue of S and λ_n the greatest, the proof relies on the relations

$$\sqrt{\lambda_1}\|\Delta x\| \leq \|\Delta x\|_S \leq \sqrt{\lambda_n}\|\Delta x\| \quad \text{for all } \Delta x,$$

as follow easily from the spectral decomposition of S.

To prove that $S^{-1}g(x)$ is the gradient of f at x w.r.t. $\langle\,,\,\rangle_S$, we wish to show

$$\limsup_{\|\Delta x\|_S \to 0} \frac{|f(x + \Delta x) - f(x) - \langle S^{-1}g(x), \Delta x \rangle_S|}{\|\Delta x\|_S} = 0.$$

However, noting that for all v,

$$\langle S^{-1}g(x), v \rangle_S = \langle S^{-1}g(x), Sv \rangle = \langle SS^{-1}g(x), v \rangle = \langle g(x), v \rangle,$$

we have

$$\limsup_{\|\Delta x\|_S \to 0} \frac{|f(x + \Delta x) - f(x) - \langle S^{-1}g(x), \Delta x \rangle_S|}{\|\Delta x\|_S}$$

$$= \limsup_{\|\Delta x\|_S \to 0} \frac{|f(x + \Delta x) - f(x) - \langle g(x), \Delta x \rangle|}{\|\Delta x\|_S}$$

$$\leq \frac{1}{\sqrt{\lambda_1}} \limsup_{\|\Delta x\| \to 0} \frac{|f(x + \Delta x) - f(x) - \langle g(x), \Delta x \rangle|}{\|\Delta x\|}$$

$$= 0,$$

completing the proof. □

Theorem 1.2.1 has the unsurprising consequence that the first-order approximation of f at x is independent of the inner product:

$$f(x) + \langle g(x), y - x \rangle = f(x) + \langle S^{-1}g(x), y - x \rangle_S.$$

Finally, we make a few observations that will be important for applying ipm theory to optimization problems having linear equations among the constraints.

Recall that we use $\mathbb{R}^n$ to denote a generic finite-dimensional vector space with an inner product, and hence our definition of the gradient applies to subspaces, too. In particular, if f is defined on a vector space and the domain of f intersects a subspace L, one can speak of the gradient of the function $f|_L$ obtained by restricting f to L. Substituting L for $\mathbb{R}^n$ in the definition of the gradient, we see that the gradient of $f|_L$ at $x \in D_f \cap L$ is the vector $g|_L(x) \in L$ satisfying

$$\lim_{\Delta x \in L,\, \|\Delta x\| \to 0} \frac{f(x + \Delta x) - f(x) - \langle g|_L(x), \Delta x \rangle}{\|\Delta x\|} = 0.$$

If D_f is an open subset of $\mathbb{R}^n$ and $D_f \cap L \neq \emptyset$ where L is a subspace, then for $x \in D_f \cap L$ we can refer both to the gradient $g(x) \in \mathbb{R}^n$ and to the gradient $g|_L(x) \in L$. We claim that $g|_L(x) = P_L g(x)$ where P_L is the operator projecting $\mathbb{R}^n$ orthogonally onto L. Indeed, it is readily proven that P_L is self-adjoint (e.g., use the facts that $P_L u = u$ if $u \in L$, $P_L v = 0$ if $v \in L^\perp$ and each vector w in $\mathbb{R}^n$ can be expressed $w = u + v$ where $u \in L$, $v \in L^\perp$). Consequently,

$$\begin{aligned}
0 &= \lim_{\Delta x \in \mathbb{R}^n,\, \|\Delta x\| \to 0} \frac{f(x + \Delta x) - f(x) - \langle g(x), \Delta x \rangle}{\|\Delta x\|} \\
&= \lim_{\Delta x \in L,\, \|\Delta x\| \to 0} \frac{f(x + \Delta x) - f(x) - \langle g(x), \Delta x \rangle}{\|\Delta x\|} \\
&= \lim_{\Delta x \in L,\, \|\Delta x\| \to 0} \frac{f|_L(x + \Delta x) - f|_L(x) - \langle g(x), P_L \Delta x \rangle}{\|\Delta x\|} \\
&= \lim_{\Delta x \in L,\, \|\Delta x\| \to 0} \frac{f|_L(x + \Delta x) - f|_L(x) - \langle P_L g(x), \Delta x \rangle}{\|\Delta x,\|},
\end{aligned}$$

and hence $g|_L(x) = P_L g(x)$ as claimed.

Similarly, if L' is a translate of a subspace L and $D_f \cap L' \neq \emptyset$, we can consider the restricted functional $f|_{L'}$. As the gradient is naturally defined for subspaces rather than for affine spaces like L', it is sometimes expedient to instead consider the functional with domain L defined by $v \mapsto f|_{L'}(v + x)$, where x is a fixed point in L'. In this way, facts about gradients can be applied to analyzing $f|_{L'}$. We sometimes state results for functionals restricted to affine spaces while giving only proofs for subspaces, leaving the remaining details to the reader.

1.3 Hessians

The functional f is said to be *twice differentiable* at $x \in D_f$ if $f \in \mathcal{C}^1$ and there exists a linear operator $H(x) : \mathbb{R}^n \to \mathbb{R}^n$ satisfying

$$\lim_{\|\Delta x\| \to 0} \frac{\|g(x + \Delta x) - g(x) - H(x)\Delta x\|}{\|\Delta x\|} = 0.$$

If it exists, $H(x)$ is said to be the *Hessian* of f at x w.r.t. $\langle\,,\,\rangle$.

If $\langle\,,\,\rangle$ is the dot product and $H(x)$ is written as a matrix, the (i, j) entry of $H(x)$ is

$$\frac{\partial^2 f}{\partial x_j \partial x_i}.$$

The *second-order approximation of f at x* is the quadratic functional

$$y \mapsto f(x) + \langle g(x), y - x \rangle + \tfrac{1}{2} \langle (y - x), H(x)(y - x) \rangle.$$

If f is twice differentiable at each $x \in D_f$, then f is said to be *twice differentiable*. Henceforth, assume f is twice differentiable.

If the function $x \mapsto H(x)$ is continuous at x (w.r.t. the operator-norm topology or, equivalently, any norm topology on the vector space of linear operators from $\mathbb{R}^n$ to $\mathbb{R}^n$), then $H(x)$ is self-adjoint. If the function $x \mapsto H(x)$ is continuous at each $x \in D_f$, then f is said to be *twice continuously differentiable* and one writes $f \in \mathcal{C}^2$.

The assumption of twice continuous differentiability, as opposed to mere twice differentiability, is often made in optimization to guarantee self-adjointness of the Hessian. In this book, functionals are always assumed to be twice continuously differentiable.

If the inner product is the dot product and the Hessian is expressed as a matrix, self-adjointness is equivalent to the matrix being symmetric, that is,

$$\frac{\partial^2 f}{\partial x_i \partial x_j} = \frac{\partial^2 f}{\partial x_j \partial x_i}.$$

(If the Hessian matrix does not vary continuously in x, the order in which the partials are taken can matter, resulting in a nonsymmetric matrix.)

To illustrate the definition of the Hessian, we again consider the functional

$$f(X) := -\ln \det(X)$$

with domain $\mathbb{S}_{++}^{n \times n}$, the set of all pd matrices in $\mathbb{S}^{n \times n}$. We saw that $g(X) = -X^{-1}$. We claim that $H(X)$ is the linear operator given by

$$\Delta X \mapsto X^{-1}(\Delta X)X^{-1}.$$

This can be proven by relying on the fact that if $\|\Delta X\|$ is sufficiently small, then

$$(X + \Delta X)^{-1} = X^{-1} \sum_{k=0}^{\infty} [-(\Delta X)X^{-1}]^k,$$

and hence

$$g(X + \Delta X) - g(X) - X^{-1}(\Delta X)X^{-1} = -X^{-1}\sum_{k=2}^{\infty}[-(\Delta X)X^{-1}]^k.$$

For then, from the submultiplicativity of the Frobenius norm,

$$\limsup_{\|\Delta X\| \to 0} \frac{\|g(X + \Delta X) - g(X) - X^{-1}(\Delta X)X^{-1}\|}{\|\Delta X\|}$$
$$\leq \limsup_{\|\Delta X\| \to 0} \frac{\|\Delta X\|^2 \|X^{-1}\|^3 \sum_{k=0}^{\infty}(\|(\Delta X)\| \|X^{-1}\|)^k}{\|\Delta X\|}$$
$$= 0,$$

establishing that the Hessian is as claimed.

Since $f(X) := -\ln \det(X)$ is twice continuously differentiable (in fact, analytic), the Hessian $H(X)$ is self-adjoint and hence has a spectral decomposition, i.e., has an orthonormal basis of eigenvectors, where "orthonormal" is w.r.t. the trace product. Indeed, letting $v_1, \ldots, v_n$ be an orthonormal basis of eigenvectors for the symmetric matrix X (here, "orthonormal" refers to the dot product on $\mathbb{R}^n$), and letting $\lambda_1, \ldots, \lambda_n$ be the eigenvalues, it is readily verified that an orthonormal basis of eigenvectors for the operator $H(X)$ is given by the set of matrices $\{M_{ij} : 1 \leq i \leq j \leq n\}$ where

$$M_{ij} := \begin{cases} v_i v_i^T & \text{if } i = j, \\ \frac{1}{\sqrt{2}}(v_i v_j^T + v_j v_i^T) & \text{if } i \neq j. \end{cases}$$

Moreover, it is readily verified that the eigenvalue for M_{ij} is $\lambda_i \lambda_j$.

The property of being twice continuously differentiable does not depend on the inner product, whereas the Hessian most certainly does depend on the inner product. The dependence of the Hessian on the inner product is made explicit in the following theorem. The proof of the theorem is similar to the proof of Theorem 1.2.1 and hence is left to the reader.

Theorem 1.3.1. *If S is pd and f is twice differentiable at x, then the Hessian of f at x w.r.t. $\langle \, , \, \rangle_S$ is $S^{-1}H(x)$.*

The lack of symmetry in the expression $S^{-1}H(x)$ might puzzle readers who are thinking of Hessians as being symmetric matrices. We remark that if one chose a basis for $\mathbb{R}^n$ which is orthonormal w.r.t. $\langle \, , \, \rangle_S$ and expressed the operator $S^{-1}H(x)$ as a matrix in terms of the resulting coordinate system, then the matrix would be symmetric. (Again we emphasize that in developing ipm theory, it is best not to think in terms of coordinate systems.)

Theorems 1.2.1 and 1.3.1 have the unsurprising consequence that the second-order approximation of f at x is independent of the inner product:

$$f(x) + \langle g(x), y - x \rangle + \tfrac{1}{2} \langle (y - x), H(x)(y - x) \rangle$$
$$= f(x) + \langle S^{-1}g(x), y - x \rangle_S + \tfrac{1}{2} \langle (y - x), S^{-1}H(x)(y - x) \rangle_S.$$

Finally, we make an observation regarding Hessians and subspaces. It is straightforward to prove that if L is a subspace of $\mathbb{R}^n$ and f satisfies $D_f \cap L \neq \emptyset$, then the Hessian of $f|_L$ at $x \in L$—an operator from L to L—is given by $H|_L(x) = P_L H(x) \ (= P_L H(x) P_L)$ where $P|_L$ is the operator which orthogonally projects $\mathbb{R}^n$ onto L. That is, when one applies $H|_L(x)$ to a vector $v \in L$, one obtains the vector $P_L H(x) v$.

1.4 Convexity

Recall that a set $S \subseteq \mathbb{R}^n$ is said to be *convex* if whenever $x, y \in S$ and $0 \leq t \leq 1$ we have $x + t(y - x) \in S$.

Recall that a functional f is said to be *convex* if D_f is convex and if whenever $x, y \in D_f$ and $0 \leq t \leq 1$, we have

$$f(x + t(y - x)) \leq f(x) + t(f(y) - f(x)).$$

If the inequality is strict whenever $0 < t < 1$ and $x \neq y$, then f is said to be *strictly convex*.

The minimizers of a convex functional form a convex set. A strictly convex functional has at most one minimizer.

Henceforth, we assume $f \in \mathcal{C}^2$ and we assume D_f is an open, convex set.

If f is a univariate functional, we know from calculus that f is convex iff $f''(x) \geq 0$ for all $x \in D_f$. Similarly, if $f''(x) > 0$ for all $x \in D_f$, then f is strictly convex. The following standard theorem generalizes these facts.

Theorem 1.4.1. *The functional f is convex iff $H(x)$ is psd for all $x \in D_f$. If $H(x)$ is pd for all $x \in D_f$, then f is strictly convex.*

The following elementary proposition, which is relied on in the proof of Theorem 1.4.1, is fundamental in this book. It does not assume convexity of f.

Proposition 1.4.2. *Assume $x, y \in D_f$ and define a univariate functional $\phi : [0, 1] \to \mathbb{R}$ by*

$$\phi(t) := f(x + t(y - x)).$$

Then

$$\phi'(t) = \langle g(x + t(y - x)), y - x \rangle$$

and

$$\phi''(t) = \langle y - x, H(x + t(y - x))(y - x) \rangle.$$

Proof. Fix t and let $u = x + t(y - x)$. We wish to prove

$$\phi'(t) = \langle g(u), y - x \rangle \quad \text{and} \quad \phi''(t) = \langle y - x, H(u)(y - x) \rangle.$$

To prove $\phi'(t) = \langle g(u), y - x \rangle$ it suffices to show

$$\limsup_{s \to 0} \left| \frac{\phi(t + s) - \phi(t) - s \langle g(u), y - x \rangle}{s} \right| = 0.$$

However, noting $\phi(t) = f(u)$ and $\phi(t+s) = f(u+s(y-x))$, we have

$$\limsup_{s \to 0} \left| \frac{\phi(t+s) - \phi(t) - s\langle g(u), y-x \rangle}{s} \right|$$
$$= \limsup_{s \to 0} \left| \frac{f(u+s(y-x)) - f(u) - \langle g(u), s(y-x) \rangle}{s} \right|$$
$$= \|y-x\| \limsup_{\|s(y-x)\| \to 0} \frac{|f(u+s(y-x)) - f(u) - \langle g(u), s(y-x) \rangle|}{\|s(y-x)\|}$$
$$= 0,$$

the final equality by definition of $g(u)$.

Similarly, to prove $\phi''(t) = \langle y-x, H(u)(y-x) \rangle$, it suffices to show

$$\limsup_{s \to 0} \left| \frac{\phi'(t+s) - \phi'(t) - s\langle y-x, H(u)(y-x) \rangle}{s} \right| = 0.$$

However, since we now know

$$\phi'(t) = \langle g(u), y-x \rangle \quad \text{and} \quad \phi'(t+s) = \langle g(u+s(y-x)), y-x \rangle,$$

we have

$$\limsup_{s \to 0} \left| \frac{\phi'(t+s) - \phi'(t) - s\langle y-x, H(u)(y-x) \rangle}{s} \right|$$
$$= \limsup_{s \to 0} \left| \frac{\langle g(u+s(y-x)) - g(u) - H(u)s(y-x), y-x \rangle}{s} \right|$$
$$\leq \limsup_{s \to 0} \frac{\|g(u+s(y-x)) - g(u) - H(u)s(y-x)\| \, \|y-x\|}{|s|}$$
$$= \|y-x\|^2 \limsup_{\|s(y-x)\| \to 0} \frac{\|g(u+s(y-x)) - g(u) - H(u)s(y-x)\|}{\|s(y-x)\|}$$
$$= 0,$$

where the inequality is by Cauchy–Schwarz and the final equality is by definition of $H(u)$. □

Proof of Theorem 1.4.1. We first show that if the Hessian is psd everywhere on D_f, then f is convex. So assume the Hessian is psd everywhere on D_f.

Assume x and y are arbitrary points in D_f. We wish to show that if t satisfies $0 \leq t \leq 1$, then

$$f(x + t(y-x)) \leq f(x) + t(f(y) - f(x)). \tag{1.2}$$

Consider the univariate functional ϕ defined by

$$\phi(t) := f(x + t(y-x)).$$

Observe that (1.2) is equivalent to

$$\phi(t) \leq \phi(0) + t(\phi(1) - \phi(0)),$$

1.4. Convexity

an inequality that is certainly valid if ϕ is convex on the interval $[0, 1]$. Hence to prove (1.2) it suffices to prove $\phi''(t) \geq 0$ for all $0 \leq t \leq 1$. However, Proposition 1.4.2 implies

$$\phi''(t) = \langle y - x, H(x + t(y - x))(y - x) \rangle \geq 0,$$

the inequality because $H(x + t(y - x))$ is psd.

The proof that f is strictly convex if the Hessian is everywhere pd on D_f is similar and hence is left to the reader.

To conclude the proof, it suffices to show that if $H(x)$ is not psd for some x, then f is not convex. If $H(x)$ is not psd, then $H(x)$ has an eigenvector v with negative eigenvalue λ. To show f is not convex, it suffices to show that the functional $\phi(t) := f(x + tv)$ is not convex. To show ϕ is not convex, it suffices to show $\phi''(0) < 0$. This is straightforward, again relying on Proposition 1.4.2. $\square$

It was asserted earlier that, roughly speaking, the gradient $g(x)$ points in the direction for which the functional output increases the fastest per unit distance traveled, and the magnitude $\|g(x)\|$ equals the amount the functional will change per unit distance traveled in that direction. Proposition 1.4.2 provides the means to make this rigorous. Indeed, choose an arbitrary direction v of unit length. The initial rate of change in the output of f as one moves from x to $x + v$ in unit time is given by $\phi'(0)$ where

$$\phi(t) := f(x + tv).$$

Note that Proposition 1.4.2 implies

$$\phi'(0) = \langle g(x), v \rangle, \tag{1.3}$$

and hence, by Cauchy–Schwarz and $\|v\| = 1$, we have

$$-\|g(x)\| \leq \phi'(0) \leq \|g(x)\|.$$

So the initial rate of change cannot exceed $\|g(x)\|$ in magnitude, regardless of which direction v of unit length is chosen. However, assuming $g(x) \neq 0$, if one chooses the direction

$$v = \frac{1}{\|g(x)\|} g(x),$$

then (1.3) implies $\phi'(0) = \|g(x)\|$.

We mention that a point z minimizes a convex functional f iff $g(z) = 0$. (For the "if," assume $g(z) = 0$. For $y \in D_f$, consider the univariate functional $\phi(t) := f(z + t(y - z))$. By Proposition 1.4.2, $\phi'(0) = 0$. Since ϕ is convex, we know from univariate calculus that 0 minimizes ϕ. In particular, $\phi(0) \leq \phi(1)$, that is, $f(z) \leq f(y)$. For the "only if," assume $g(z) \neq 0$ and consider $y := x - tg(z)$ for small $t > 0$.)

As with the two preceding sections, we close this one with a discussion of subspaces. If L is a subspace of $\mathbb{R}^n$ and $D_f \cap L \neq \emptyset$, we know the gradient of $f|_L$ to be $P_L g$, the orthogonal projection of g onto L. Thus, $z \in L$ solves the linearly constrained convex optimization problem

$$\begin{aligned} \min \quad & f(x) \\ \text{s.t.} \quad & x \in L \end{aligned}$$

iff $P_L g(z) = 0$, that is, iff $g(z)$ is orthogonal to L. In particular, if L is the nullspace of a linear operator A, then z solves the optimization problem iff $g(z) = A^* y$ for some y. The same holds when L is replaced by a translate of L, that is, when L is replaced by an affine space.

Theorem 1.4.3. *If f is convex and A is a linear operator, then $z \in D_f$ solves the linearly constrained optimization problem*

$$\begin{aligned} \min \quad & f(x) \\ \text{s.t.} \quad & Ax = b \end{aligned}$$

iff $Az = b$ and $g(z) = A^* y$ for some y.

1.5 Fundamental Theorems of Calculus

We continue to assume $f \in C^2$ and D_f is an open, convex set.

The following theorem generalizes the fundamental theorem of calculus.

Theorem 1.5.1. *If $x, y \in D_f$, then*

$$f(y) - f(x) = \int_0^1 \langle g(x + t(y - x)), y - x \rangle \, dt.$$

Proof. Consider the univariate functional $\phi(t) := f(x + t(y - x))$. The fundamental theorem of calculus asserts

$$\phi(1) - \phi(0) = \int_0^1 \phi'(t) \, dt.$$

Since $\phi(1) = f(y)$, $\phi(0) = f(x)$ and, by Proposition 1.4.2,

$$\phi'(t) = \langle g(x + t(y - x)), y - x \rangle,$$

the proof is complete. □

In a similar vein, we have the following proposition.

Proposition 1.5.2. *If $x, y \in D_f$, then*

$$f(y) = f(x) + \langle g(x), y - x \rangle \qquad (1.4)$$
$$+ \int_0^1 \langle g(x + t(y - x)) - g(x), y - x \rangle \, dt$$

and

$$f(y) = f(x) + \langle g(x), y - x \rangle + \tfrac{1}{2} \langle y - x, H(x)(y - x) \rangle \qquad (1.5)$$
$$+ \int_0^1 \int_0^t \langle y - x, [H(x + s(y - x)) - H(x)](y - x) \rangle \, ds \, dt.$$

1.5. Fundamental Theorems of Calculus

Proof. Again considering the univariate functional $\phi(t) := f(x+t(y-x))$, the fundamental theorem of calculus implies

$$\phi(1) = \phi(0) + \phi'(0) + \int_0^1 \phi'(t) - \phi'(0)\, dt \tag{1.6}$$

and

$$\phi(1) = \phi(0) + \phi'(0) + \tfrac{1}{2}\phi''(0) + \int_0^1 \int_0^t \phi''(s) - \phi''(0)\, ds\, dt. \tag{1.7}$$

Using Proposition 1.4.2 to make the obvious substitutions, (1.6) yields (1.4), whereas (1.7) yields (1.5). □

Proposition 1.5.2 provides the means to bound the error in the first- and second-order approximations of f.

Corollary 1.5.3. *If $x, y \in D_f$, then*

$$|f(y) - f(x) - \langle g(x), y - x \rangle| \leq \|y - x\| \int_0^1 \|g(x + t(y - x)) - g(x)\|\, dt$$

and

$$|f(y) - f(x) - \langle g(x), y - x \rangle - \tfrac{1}{2}\langle y - x, H(x)(y - x)\rangle|$$
$$\leq \|y - x\|^2 \int_0^1 \int_0^t \|H(x + s(y - x)) - H(x)\|\, ds\, dt.$$

Relying on continuity of g and H, observe that the error in the first-order approximation is $o(\|y - x\|)$ (i.e., tends to zero faster than $\|y - x\|$), whereas the error in the second-order approximation is $o(\|y - x\|^2)$.

Theorem 1.5.1 gives a fundamental theorem of calculus for a functional f. It will be necessary to have an analogous theorem for g, a theorem which expresses the difference $g(y) - g(x)$ as an integral involving the Hessian. To keep our development coordinate-free, we introduce the following definition:

The univariate function $t \mapsto v(t) \in \mathbb{R}^n$, with domain $[a, b]$, is said to be *integrable* if there exists a vector u such that

$$\langle u, w \rangle = \int_a^b \langle v(t), w \rangle\, dt \quad \text{for all } w \in \mathbb{R}^n.$$

If it exists, the vector u is uniquely determined (as is not difficult to prove) and is called the *integral* of the function $v(t)$. One uses the notation $\int_a^b v(t)\, dt$ to represent this vector.

Although the definition of the integral is phrased in terms of the inner product $\langle\,,\,\rangle$, it is independent of the inner product. Indeed, if u is the integral as defined by $\langle\,,\,\rangle$ and if S is pd, then for all vectors w,

$$\langle u, w \rangle_S = \langle u, Sw \rangle$$
$$= \int_a^b \langle v(t), Sw \rangle\, dt$$
$$= \int_a^b \langle v(t), w \rangle_S\, dt.$$

The following are two useful, elementary propositions.

Proposition 1.5.4. *If the univariate function $t \mapsto v(t) \in \mathbb{R}^n$, with domain $[a, b]$, is integrable, then*
$$\left\| \int_a^b v(t)\, dt \right\| \leq \int_a^b \|v(t)\|\, dt.$$

Proof. Let $u := \int_a^b v(t)\, dt$. By definition of the integral, for all w we have
$$\langle u, w \rangle = \int_a^b \langle v(t), w \rangle\, dt.$$

In particular, choosing $w = u$ gives
$$\|u\|^2 = \int_a^b \langle v(t), u \rangle\, dt. \tag{1.8}$$

However,
$$\int_a^b \langle v(t), u \rangle\, dt \leq \left| \int_a^b \langle v(t), u \rangle\, dt \right|$$
$$\leq \int_a^b |\langle v(t), u \rangle|\, dt$$
$$\leq \int_a^b \|v(t)\|\, \|u\|\, dt$$
$$= \|u\| \int_a^b \|v(t)\|\, dt. \tag{1.9}$$

Combining (1.8) and (1.9) gives
$$\|u\|^2 \leq \|u\| \int_a^b \|v(t)\|\, dt.$$

Since $\|u\| = \|\int_a^b v(t)\, dt\|$, the proof is complete. $\square$

Proposition 1.5.5. *If the univariate function $t \mapsto v(t) \in \mathbb{R}^n$, with domain $[a, b]$, is integrable and if $A : \mathbb{R}^n \to \mathbb{R}^m$ is a linear operator, then the function $t \mapsto Av(t)$ is integrable and*
$$\int_a^b Av(t)\, dt = A \int_a^b v(t)\, dt.$$

Proof. Observe that for all $w \in \mathbb{R}^m$ we have
$$\left\langle A \int_a^b v(t)\, dt, w \right\rangle = \left\langle \int_a^b v(t)\, dt, A^* w \right\rangle$$
$$= \int_a^b \langle v(t), A^* w \rangle\, dt$$
$$= \int_a^b \langle Av(t), w \rangle\, dt,$$

1.5. Fundamental Theorems of Calculus

where the second equality is by definition of $\int_a^b v(t)\,dt$. □

Next is the fundamental theorem of calculus for the gradient.

Theorem 1.5.6. *If $x, y \in D_f$, then*

$$g(y) - g(x) = \int_0^1 H(x + t(y - x))(y - x)\,dt.$$

Proof. By definition of the integral, we wish to prove that for all w,

$$\langle g(y) - g(x), w \rangle = \int_0^1 \langle H(x + t(y - x))(y - x), w \rangle\,dt. \tag{1.10}$$

Fix arbitrary w and consider the functional

$$\phi(t) := \langle g(x + t(y - x)), w \rangle.$$

The fundamental theorem of calculus asserts

$$\phi(1) - \phi(0) = \int_0^1 \phi'(t)\,dt,$$

which, by definition of ϕ, is equivalent to

$$\langle g(y) - g(x), w \rangle = \int_0^1 \phi'(t)\,dt. \tag{1.11}$$

Comparing (1.11) with (1.10), we see that to prove (1.10) it suffices to show for arbitrary $0 \le t \le 1$ that

$$\phi'(t) = \langle H(u)(y - x), w \rangle, \tag{1.12}$$

where

$$u := x + t(y - x).$$

Toward proving (1.12), recall that $H(u)$ is the unique operator satisfying

$$0 = \lim_{\|\Delta u\| \to 0} \frac{\|g(u + \Delta u) - g(u) - H(u)\Delta u\|}{\|\Delta u\|}. \tag{1.13}$$

Thinking of Δu as being $s(y - x)$ where $s \ne 0$, it follows from (1.13) that

$$0 = \lim_{s \to 0} \frac{\|g(u + s(y - x)) - g(u) - sH(u)(y - x)\|}{s}. \tag{1.14}$$

Since, by Cauchy–Schwarz,

$$\|g(u + s(y - x)) - g(u) - sH(u)(y - x)\|\,\|w\|$$
$$\ge |\langle g(u + s(y - x)) - g(u) - sH(u)(y - x), w \rangle|,$$

equation (1.14) implies

$$0 = \lim_{s \to 0} \frac{\langle g(u + s(y - x)) - g(u) - sH(u)(y - x), w \rangle}{s}.$$

Since
$$\phi(t+s) = \langle g(u+s(y-x)), w\rangle \quad \text{and} \quad \phi(t) = \langle g(u), w\rangle,$$
we thus have
$$0 = \lim_{s \to 0} \frac{\phi(t+s) - \phi(t) - s\langle H(u)(y-x), w\rangle}{s},$$
from which it is immediate that $\langle H(u)(y-x), w\rangle = \phi'(t)$. Thus, (1.12) is established and the proof is complete. □

Proposition 1.5.7. *If $x, y \in D_f$, then*
$$g(y) = g(x) + H(x)(y-x) + \int_0^1 [H(x+t(y-x)) - H(x)](y-x) \, dt.$$

Proof. The proof is a simple consequence of Theorem 1.5.6 and
$$\int_0^1 H(x)(y-x) \, dt = H(x)(y-x),$$
an identity which is trivially verified. □

Corollary 1.5.8. *If $x, y \in D_f$, then*
$$\|g(y) - g(x) - H(x)(y-x)\| \le \|y-x\| \int_0^1 \|H(x+t(y-x)) - H(x)\| \, dt.$$

1.6 Newton's Method

We continue to assume $f \in C^2$ and D_f is an open, convex set.

In optimization, Newton's method is an algorithm for minimizing functionals. The idea behind the algorithm is simple. Given a point x in the domain of a functional f, where f is to be minimized, one replaces f by the second-order approximation at x and minimizes the approximation to obtain a new point, x_+. One repeats this procedure with x_+ in place of x, and so on, generating a sequence of points which, under certain conditions, converges rapidly to a minimizer of f.

For $x \in D_f$, we denote the second-order—or "quadratic"—approximation of f at x by
$$q_x(y) := f(x) + \langle g(x), y-x\rangle + \tfrac{1}{2}\langle y-x, H(x)(y-x)\rangle.$$
The domain of q_x is all of $\mathbb{R}^n$.

Proposition 1.6.1. *The gradient of q_x at y is $g(x) + H(x)(y-x)$ and the Hessian is $H(x)$ (regardless of y).*

Proof. Using the self-adjointness of $H(x)$, it is easily established that
$$q_x(y + \Delta y) - q_x(y) - \langle g(x) + H(x)(y-x), \Delta y\rangle = \tfrac{1}{2}\langle \Delta y, H(x)\Delta y\rangle.$$

1.6. Newton's Method

Proving the gradient is as asserted is thus equivalent to proving

$$\lim_{\|\Delta y\| \to 0} \frac{\langle \Delta y, H(x)\Delta y \rangle}{\|\Delta y\|} = 0,$$

an easily established identity.

Having proven the gradient is as asserted, it is simple to prove the Hessian is as asserted. $\square$

Henceforth, assume $H(x)$ is pd. Thus, q_x is strictly convex and is minimized by the point x_+ satisfying $g(x) + H(x)(x_+ - x) = 0$, that is, q_x is minimized by the point

$$x_+ := x - H(x)^{-1}g(x).$$

The "Newton step at x" is defined to be the difference

$$n(x) := x_+ - x = -H(x)^{-1}g(x).$$

Newton's method steps from x to $x + n(x)$.

We emphasize that x_+ is the root of the map $y \mapsto g(x) + H(x)(y - x)$, the linear approximation at x of the gradient map $y \mapsto g(y)$. Although it is customary in optimization to present Newton's method as the algorithm which attempts to approximate a minimizer of f by minimizing the quadratic approximation $y \mapsto q_x(y)$ of f, one can equivalently present it as the algorithm which attempts to approximate a solution to the nonlinear equations $g(y) = 0$ by computing the root of the linear approximation $y \mapsto g(x) + H(x)(y - x)$ of g. The latter viewpoint dovetails with the use of Newton's method in applied mathematics beyond optimization; it is the algorithm which attempts to approximate a root to a general nonlinear map $g : \mathbb{R}^n \to \mathbb{R}^n$ by computing the root of the linear approximation to the map.

We know the second-order approximation is independent of the inner product. Consequently, so is Newton's method. More explicitly, in the inner product $\langle\ ,\ \rangle_S$, the gradient of f at x is $S^{-1}g(x)$, the Hessian is $S^{-1}H(x)$, and so the Newton step is

$$-(S^{-1}H(x))^{-1}S^{-1}g(x) = -H(x)^{-1}g(x).$$

The Newton step is independent from the inner product.

The following theorem is the main tool for analyzing the progress of Newton's method.

Theorem 1.6.2. *If z minimizes f and $H(x)$ is invertible, then*

$$\|x_+ - z\| \le \|x - z\|\,\|H(x)^{-1}\| \int_0^1 \|H(x + t(z - x)) - H(x)\|\, dt.$$

Proof. Noting $g(z) = 0$, we have

$$\begin{aligned}
\|x_+ - z\| &= \|x - z - H(x)^{-1}g(x)\| \\
&= \|x - z + H(x)^{-1}(g(z) - g(x))\| \\
&= \left\| x - z + H(x)^{-1} \int_0^1 H(x + t(z - x))(z - x)\, dt \right\| \\
&= \left\| H(x)^{-1} \int_0^1 [H(x + t(z - x)) - H(x)](z - x)\, dt \right\| \\
&\le \|x - z\|\,\|H(x)^{-1}\| \int_0^1 \|H(x + t(z - x)) - H(x)\|\, dt. \quad \square
\end{aligned}$$

Invoking the assumed continuity of the Hessian, the theorem is seen to imply that if $H(z)$ is invertible and x is sufficiently close to z, then x_+ will be closer to z than is x.

Now we present a brief discussion of Newton's method and subspaces, as will be important when we consider applications of ipm theory to optimization problems having linear equations among the constraints. Assume L is a subspace of $\mathbb{R}^n$ and $x \in L \cap D_f$. Let $n|_L(x)$ denote the Newton step for $f|_L$ at x. Since the gradient of $f|_L$ at x is $P_L g(x)$ and the Hessian is $P_L H(x)$, the Newton step $n|_L(x)$ is the vector in L solving

$$P_L H(x) n|_L(x) = -P_L g(x);$$

that is, $n|_L(x)$ is the vector in L for which $H(x) n|_L(x) + g(x)$ is orthogonal to L. In particular, if L is the nullspace of a linear operator $A : \mathbb{R}^n \to \mathbb{R}^m$, then $n|_L(x)$ is the vector in $\mathbb{R}^n$ for which there exists $y \in \mathbb{R}^m$ satisfying

$$H(x) n|_L(x) + g(x) = A^* y,$$
$$A n|_L(x) = 0.$$

Computing $n|_L(x)$ (and y) can thus be accomplished by solving a system of $m+n$ equations in $m+n$ variables.

If $H(x)^{-1}$ is readily computed (as it is for functionals f used in ipm's), the size of the system of linear equations to be solved can easily be reduced to m variables. One solves the linear system in the variables y,

$$A H(x)^{-1} A^* y = A H(x)^{-1} g(x),$$

and then computes

$$n|_L(x) = H(x)^{-1} (A^* y - g(x)).$$

In closing this section we remark that the error bound given by Theorem 1.6.2 can of course be applied to $f|_L$. Assuming z' minimizes $f|_L$ (trivially, the minimizer z of f over all of D_f satisfies $z = z'$ iff $z \in L$), we obtain

$$\|(x + n|_L(x)) - z'\| \leq \|x - z'\| \, \|H|_L(x)^{-1}\| \int_0^1 \|H|_L(x + t(z' - x)) - H|_L(x)\| \, dt.$$

However, since $H|_L(y) = P_L H(y) P_L$ for all $y \in D_f$, it is easily proven that

$$\|H|_L(x)^{-1}\| \leq \|H(x)^{-1}\| \quad \text{and} \quad \|H|_L(y) - H|_L(x)\| \leq \|H(y) - H(x)\|.$$

Hence,

$$\|(x + n|_L(x)) - z'\| \leq \|x - z'\| \, \|H(x)^{-1}\| \int_0^1 \|H(x + t(z' - x)) - H(x)\| \, dt.$$

The Hessian on the right being H rather than $H|_L$ makes for less cumbersome applications of the inequality.

Chapter 2
Basic Interior-Point Method Theory

Throughout this chapter, unless otherwise stated, f refers to a functional having at least the following properties: D_f is open and convex, $f \in \mathcal{C}^2$, and $H(x)$ is pd for all $x \in D_f$. In particular, f is strictly convex.

2.1 Intrinsic Inner Products

The functional f gives rise to a family of inner products, an inner product for each point $x \in D_f$:
$$\langle u, v \rangle_{H(x)} := \langle u, H(x)v \rangle.$$

These inner products vary continuously with x. In particular, given $\epsilon > 0$, there exists a neighborhood of x consisting of points y with the property that for all vectors $v \neq 0$,
$$1 - \epsilon < \frac{\|v\|_{H(y)}}{\|v\|_{H(x)}} < 1 + \epsilon.$$

We often refer to the inner product $\langle\ ,\ \rangle_{H(x)}$ as the *local inner product (at x)*.

In the inner product $\langle\ ,\ \rangle_{H(x)}$, the gradient at y is $H(x)^{-1}g(y)$ and the Hessian is $H(x)^{-1}H(y)$. In particular, the gradient at x is $-n(x)$, the negative of the Newton step, and the Hessian is I, the identity. Thus, in the local inner product, Newton's method coincides with the "method of steepest descent," i.e., Newton's method coincides with the algorithm which attempts to minimize f by moving in the direction given by the negative of the gradient. (Whereas Newton's method is independent of inner products, the method of steepest descent is not independent because gradients are not independent.)

It appears from our definition that the local inner product potentially depends on the reference inner product $\langle\ ,\ \rangle$. In fact, the local inner product is independent of the reference inner product; indeed, if the reference inner product is changed to $\langle\ ,\ \rangle_S$, and hence the Hessian is changed to $S^{-1}H(x)$, the resulting local inner product is
$$\langle u, S^{-1}H(x)v \rangle_S = \langle u, SS^{-1}H(x)v \rangle = \langle u, H(x)v \rangle,$$

that is, the local inner product is unchanged.

The independence of the local inner products from the reference inner product shows the local inner products to be intrinsic to the functional f. For that reason, we often refer to the local inner products as the *intrinsic inner products*. To highlight the independence of the local inner products from any reference inner product, we adopt notation which avoids the Hessians of a reference. We denote the intrinsic inner product at x by $\langle\ ,\ \rangle_x$. Let $\|\ \|_x$ denote the induced norm. For $y \in D_f$, let $g_x(y)$ denote the gradient at y and let $H_x(y)$ denote the Hessian. Thus, $g_x(x) = -n(x)$ and $H_x(x) = I$. If $A : \mathbb{R}^n \to \mathbb{R}^m$ is a linear operator, let A_x^* denote its adjoint w.r.t. $\langle\ ,\ \rangle_x$. (Of course the adjoint also depends on the inner product on $\mathbb{R}^m$. That inner product will always be fixed but arbitrary, unlike the intrinsic inner products which vary with x and are not arbitrary, depending on f.)

The reader should be especially aware that we use $g_x(x)$ and $-n(x)$ interchangeably, depending on context.

A miniscule amount of the ipm literature is written in terms of the local inner products. Rather, in much of the literature, only a reference inner product is explicit, say, the dot product. There, the proofs are done by manipulating operators built from Hessians, operators like $H(x)^{-1}H(y)$ and $AH(x)^{-1}A^T$, operators we recognize as being $H_x(y)$ and AA_x^*. An advantage to working in the local inner products is that the underlying geometry becomes evident and, consequently, the operator manipulations in the proofs become less mysterious.

Observe that the quadratic approximation of f at x is

$$q_x(y) = f(x) + \langle g(x), y - x\rangle + \tfrac{1}{2}\langle y - x, H(x)(y - x)\rangle$$
$$= f(x) - \langle n(x), y - x\rangle_x + \tfrac{1}{2}\|y - x\|_x^2,$$

and its error in approximating $f(y)$ (Corollary 1.5.3) is no worse than

$$\|y - x\|_x^2 \int_0^1 \int_0^t \|I - H_x(x + s(y - x))\|_x\, ds\, dt,$$

where the latter norm is the operator norm induced by the local norm. Similarly, the progress made by Newton's method toward approximating a minimizer z (Theorem 1.6.2) is captured by the inequality

$$\|x_+ - z\|_x \le \|x - z\|_x \int_0^1 \|I - H_x(x + t(z - x))\|_x\, dt.$$

Assume L is a subspace of $\mathbb{R}^n$ and $x \in L \cap D_f$. Let $g|_{L,x}$ denote the gradient of $f|_L$ w.r.t. the inner product obtained by restricting $\langle\ ,\ \rangle_x$ to L. Likewise, let $H|_{L,x}$ denote the Hessian. Let $P_{L,x}$ denote the linear operator which projects $\mathbb{R}^n$ orthogonally onto L, orthogonally w.r.t. $\langle\ ,\ \rangle_x$. We know from sections 1.2 and 1.3 that $g|_{L,x}(y) = P_{L,x}g_x(y)$ and $H|_{L,x}(y)v = P|_{L,x}H_x(y)v$ for all $y, v \in L$.

Observe that

$$H|_{L,x}(x) \equiv P_{L,x}H_x(x) \equiv P_{L,x} \equiv I \quad \text{(the last identity is valid on } L \text{, not } \mathbb{R}^n\text{)}.$$

Consequently, the local inner product on L induced by $f|_L$ is precisely the restriction of $\langle\ ,\ \rangle_x$ to L. Thus,

$$n|_L(x) = -g|_{L,x}(x) = -P_{L,x}g_x(x) = P_{L,x}n(x).$$

That is, in the local inner product, the Newton step for $f|_L$ is the orthogonal projection of the Newton step for f.

If L is the nullspace of a surjective linear operator A, the relation

$$n|_L(x) = P_{L,x} n(x) = [I - A_x^*(AA_x^*)^{-1}A]n(x)$$

provides the means to compute $n|_L(x)$ from $n(x)$. One solves the linear system

$$AA_x^* y = -An(x)$$

and then computes

$$n|_L(x) = A_x^* y + n(x).$$

Expressed in terms of an arbitrary inner product $\langle\ ,\ \rangle$, the equations become

$$AH(x)^{-1}A^* y = AH(x)^{-1}g(x) \quad \text{and} \quad n|_L(x) = H(x)^{-1}[A^* y - g(x)],$$

precisely the equations we arrived at in §1.6 by slightly different reasoning.

2.2 Self-Concordant Functionals

2.2.1 Introduction

Let $B_x(y, r)$ denote the open ball of radius r centered at y, where the radius is measured w.r.t. $\|\ \|_x$. Let $\bar{B}_x(y, r)$ denote the closed ball.

A functional f is said to be *(strongly nondegenerate) self-concordant* if for all $x \in D_f$ we have $B_x(x, 1) \subseteq D_f$, and if whenever $y \in B_x(x, 1)$ we have

$$1 - \|y - x\|_x \leq \frac{\|v\|_y}{\|v\|_x} \leq \frac{1}{1 - \|y - x\|_x} \quad \text{for all } v \neq 0.$$

Let $\mathcal{SC}$ denote the family of functionals thus defined.

Self-concordant functionals play a central role in the general theory of ipm's, as was developed in the pioneering work of Nesterov and Nemirovskii [15]. Although our definition of strongly nondegenerate self-concordant functionals is on the surface quite different from the original definition given in [15], it is in fact equivalent except in assuming $f \in C^2$ as opposed to the ever-so-slightly stronger assumption in [15] that f is thrice continuously differentiable. The equivalence is shown in §2.5, where it is also shown that our definition can be "relaxed" in a few ways without altering the family of functionals so defined; for example, the leftmost inequality involving $\|v\|_y/\|v\|_x$ is redundant.

The term "strongly" refers to the requirement $B_x(x, 1) \subseteq D_f$. The term "nondegenerate" refers to the Hessians being pd, thereby giving the local inner products. The definition of self-concordant functionals—not necessarily strongly nondegenerate—is a natural relaxation of the above definition, only requiring the Hessians to be psd. However, it is the strongly nondegenerate self-concordant functionals that play the central role in ipm theory, and so the relaxation of the definition is best postponed until the reader has in mind a general outline of the theory.

As the parentheses in our definition indicate, for brevity we typically refer to strongly nondegenerate self-concordant functionals simply as "self-concordant functionals."

If a linear functional is added to a self-concordant functional ($x \mapsto \langle c, x \rangle + f(x)$), the resulting functional is self-concordant because the Hessians are unaffected. Similarly, if one restricts a self-concordant functional f to a subspace L (or to a translation of the subspace), one obtains a self-concordant functional, a simple consequence of the local norms for $f|_L$ being the restrictions of the local norms for f.

The primordial self-concordant barrier functional is the "logarithmic barrier function for the nonnegative orthant," having domain $D_f := \mathbb{R}^n_{++}$ (i.e., the strictly positive orthant). It is defined by $f(x) := -\sum_j \ln x_j$. Since the coordinates of vectors play a prominent role in the definition of this functional (as they do in the definition of the nonnegative orthant), to prove self-concordance it is natural to use the dot product as a reference inner product. Expressing the Hessian $H(x)$ as a matrix, one sees it is diagonal with jth diagonal entry $1/x_j^2$. Consequently, $y \in B_x(x, 1)$ is equivalent to

$$\sum_j \left(\frac{y_j - x_j}{x_j}\right)^2 < 1,$$

an inequality which is easily seen to imply $y \in D_f \ (= \mathbb{R}^n_{++})$ as required by the definition of self-concordance. Moreover, assuming $y \in B_x(x, 1)$ and v is an arbitrary vector, we have

$$\|v\|_y^2 = \sum_j \left(\frac{v_j}{y_j}\right)^2$$

$$= \sum_j \left(\frac{v_j}{x_j}\right)^2 \left(\frac{x_j}{y_j}\right)^2$$

$$\leq \|v\|_x^2 \max_j \left(\frac{x_j}{y_j}\right)^2.$$

Since
$$\tfrac{y_j}{x_j} \geq 1 - |\tfrac{y_j}{x_j} - 1| \geq 1 - \|y - x\|_x,$$

the rightmost inequality on $\|v\|_y / \|v\|_x$ in the definition of self-concordance is proven. The leftmost inequality is proven similarly. Thus, the barrier function for the nonnegative orthant is indeed self-concordant.

For an LP
$$\begin{aligned} \min \quad & c \cdot x \\ \text{s.t.} \quad & Ax = b, \\ & x \geq 0, \end{aligned}$$

the most important self-concordant functionals are those of the form

$$\eta c \cdot x + f|_L(x),$$

where $\eta \geq 0$ is a fixed constant, f is the logarithmic barrier function for the nonnegative orthant, and $L := \{x : Ax = b\}$. (Of course L is an affine space, a translation of a subspace; L is a subspace iff $b = 0$. For a brief discussion of how our results for functionals

2.2. Self-Concordant Functionals

restricted to subspaces readily translate to functionals restricted to affine spaces, recall the last paragraph of §1.2.)

Another important self-concordant functional is the "logarithmic barrier function for the cone of psd matrices" in $\mathbb{S}^{n \times n}$. This is the functional defined by $f(X) := -\ln \det(X)$, having domain $\mathbb{S}_{++}^{n \times n}$ (i.e., the pd matrices in $\mathbb{S}^{n \times n}$). To prove self-concordance, it is natural to rely on the trace product, for which we know $H(X)\Delta X = X^{-1}(\Delta X)X^{-1}$. For arbitrary $Y \in \mathbb{S}^{n \times n}$, keeping in mind that the trace of a matrix depends only on the eigenvalues, we have

$$\begin{aligned}
\|Y - X\|_X^2 &= (Y - X) \circ (X^{-1}(Y - X)X^{-1}) \\
&= \text{trace}((Y - X)X^{-1}(Y - X)X^{-1}) \\
&= \text{trace}(X^{-1/2}(Y - X)X^{-1}(Y - X)X^{-1/2}) \\
&= \sum_j (1 - \lambda_j)^2,
\end{aligned}$$

where $\lambda_1 \leq \cdots \leq \lambda_n$ are the eigenvalues of $X^{-1/2}YX^{-1/2}$. Assuming $\|Y - X\|_X < 1$, all of the values λ_j are thus positive, and hence $X^{-1/2}YX^{-1/2}$ is pd, which is easily seen to be equivalent to Y being pd. Consequently, if $\|Y - X\|_X < 1$, then $Y \in D_f \, (= \mathbb{S}_{++}^{n \times n})$, as required by the definition of self-concordance.

Assuming $Y \in B_X(X, 1)$ and $V \in \mathbb{S}^{n \times n}$, let

$$S_1 := X^{-1/2} V X^{-1/2} \quad \text{and} \quad S_2 := X^{1/2} Y^{-1} X^{1/2}.$$

Note that $\|S_1\| = \|V\|_X$ and the eigenvalues of S_2 are $0 < 1/\lambda_n \leq \cdots \leq 1/\lambda_1$. In establishing the bounds for $\|V\|_Y/\|V\|_X$ in the definition of self-concordance, we rely on the inequality

$$\|S_2^{1/2} S_1 S_2^{1/2}\| \leq \tfrac{1}{\lambda_1} \|S_1\|.$$

To verify this inequality, let Q be an orthogonal matrix diagonalizing S_2 so that $QS_2^{1/2}Q^T$ is a diagonal matrix $\Lambda^{-1/2}$ with diagonal entries $1/\sqrt{\lambda_n} \leq \cdots \leq 1/\sqrt{\lambda_1}$. Since the Frobenius norm of a symmetric matrix is determined solely by the eigenvalues, it is immediate that

$$\|S_2^{1/2} S_1 S_2^{1/2}\| = \|\Lambda^{-1/2} Q S_1 Q^T \Lambda^{-1/2}\|.$$

However, as the Frobenius norm of a matrix M satisfies $\|M\| = (\sum m_{ij}^2)^{1/2}$, it is easily seen that

$$\|\Lambda^{-1/2} Q S_1 Q^T \Lambda^{-1/2}\| \leq \tfrac{1}{\lambda_1} \|Q S_1 Q^T\| = \tfrac{1}{\lambda_1} \|S_1\|,$$

where the last equality once again relies on the Frobenius norm of a symmetric matrix being determined solely by the eigenvalues.

We have

$$\begin{aligned}
\|V\|_Y^2 &= V \circ Y^{-1} V Y^{-1} \\
&= \text{trace}(V Y^{-1} V Y^{-1}) \\
&= \text{trace}((S_1 S_2)^2) \\
&= \text{trace}((S_2^{1/2} S_1 S_2^{1/2})^2)
\end{aligned}$$

$$= (S_2^{1/2} S_1 S_2^{1/2}) \circ (S_2^{1/2} S_1 S_2^{1/2})$$
$$= \|S_2^{1/2} S_1 S_2^{1/2}\|^2$$
$$\leq \tfrac{1}{\lambda_1^2} \|S_1\|^2$$
$$= \tfrac{1}{\lambda_1^2} \|V\|_X^2.$$

Since
$$\lambda_1 \geq 1 - |1 - \lambda_1| \geq 1 - \left(\sum_{i=1}^n (1-\lambda_i)^2\right)^{1/2} = 1 - \|Y - X\|_X,$$

the rightmost inequality for $\|V\|_Y / \|V\|_X$ in the definition of self-concordance is proven. The leftmost inequality is proven similarly. Thus, the logarithmic barrier function for the cone of pd matrices is indeed self-concordant.

For an SDP
$$\begin{aligned} \min \quad & C \circ X \\ \text{s.t.} \quad & A(X) = b, \\ & X \succeq 0, \end{aligned}$$

where $A : \mathbb{S}^{n \times n} \to \mathbb{R}^m$ is a linear operator, the most important self-concordant functionals are those of the form
$$\eta C \circ X + f|_L(X),$$

where $\eta \geq 0$ is a fixed constant, f is the logarithmic barrier function for the cone of pd matrices, and $L := \{X : A(X) = b\}$.

LP can be viewed as a special case of SDP by identifying, in the obvious manner, $\mathbb{R}^n$ with the subspace in $\mathbb{S}^{n \times n}$ consisting of diagonal matrices. Then the logarithmic barrier function for the pd cone restricts to be the logarithmic barrier function for the nonnegative orthant. Thus, we were redundant in giving a proof that the logarithmic barrier function for the nonnegative orthant is indeed a self-concordant functional since the restriction of self-concordant functionals to affine spaces yield self-concordant functionals. The insight gained from the simplicity of the nonnegative orthant justifies the redundancy.

In §2.5 we show that the self-concordance of each of these two logarithmic barrier functions is a simple consequence of the original definition of self-concordance due to Nesterov and Nemirovskii [15]. (The original definition is not particularly well suited for a transparent development of the theory, but it is well suited for establishing self-concordance.)

To apply the definition of self-concordance in developing the theory, it is useful to rephrase it in terms of Hessians. (The operator norm appearing in the following theorem is the one induced by the local norm $\| \ \|_x$.)

Theorem 2.2.1. *Assume the functional f has the property that $B_x(x, 1) \subseteq D_f$ for all $x \in D_f$. Then f is self-concordant iff for all $x \in D_f$ and $y \in B_x(x, 1)$,*

$$\|H_x(y)\|_x, \|H_x(y)^{-1}\|_x \leq \frac{1}{(1 - \|y - x\|_x)^2}; \tag{2.1}$$

likewise, f is self-concordant iff

$$\|I - H_x(y)\|_x, \|I - H_x(y)^{-1}\|_x \leq \frac{1}{(1 - \|y - x\|_x)^2} - 1. \tag{2.2}$$

2.2. Self-Concordant Functionals

Proof. Let $\lambda_1 \leq \cdots \leq \lambda_n$ denote the eigenvalues of $H_x(y)$. Since

$$\max_v \frac{\|v\|_y^2}{\|v\|_x^2} = \max_v \frac{\langle v, H_x(y)v\rangle_x}{\|v\|_x^2} = \lambda_n = \|H_x(y)\|_x$$

and, similarly,

$$\min_v \frac{\|v\|_y^2}{\|v\|_x^2} = \lambda_1 = 1/\|H_x(y)^{-1}\|_x,$$

the pair of inequalities in the definition of self-concordance is equivalent to the pair (2.1). To complete the proof, we show that the pair of inequalities (2.1) is equivalent to the pair (2.2).

Since the eigenvalues of $I - H_x(y)$ are $1 - \lambda_i$, we have

$$\begin{aligned}
\|I - H_x(y)\|_x &= \max\{\lambda_n - 1, 1 - \lambda_1\} \\
&\leq \max\{\lambda_n - 1, \tfrac{1}{\lambda_1} - 1\} \\
&= \max\{\|H_x(y)\|_x - 1, \|H_x(y)^{-1}\|_x - 1\},
\end{aligned}$$

where the inequality relies on the relations $0 < \lambda_1 \leq \lambda_n$. Hence, the first inequality in (2.2) is implied by the pair of inequalities (2.1), and similarly for the second inequality in (2.2). Finally, it is immediate that the first inequality (resp., second inequality) of (2.2) implies the first inequality (resp., second inequality) of (2.1). □

Recalling that $H_x(x) = I$, it is evident from (2.2) that to assume a functional to be self-concordant is essentially to assume Lipschitz continuity of the Hessians w.r.t. the operator norms induced by the local norms.

An aside for those familar with third differentials: dividing the quantities on the left and right of (2.2) by $\|y - x\|_x$ and taking the limits supremum as y tends to x suggests, when f is thrice differentiable, that self-concordance implies the local norm of the third differential to be bounded by "2." In fact, the converse is also true, that is, a bound of "2" on the local norm of the third differential for all $x \in D_f$, together with the requirement that the local unit balls be contained in the functional domain, implies self-concordance, as we shall see in §2.5. Indeed, the original definition of self-concordance in [15] is phrased as a bound on the third differential.

2.2.2 Self-Concordancy and Newton's Method

The following theorems display the simplifying role the conditions of self-concordance play in analysis. The first theorem bounds the error of the quadratic approximation and the second guarantees progress made by Newton's method. Note the elegance of the bounds when compared to the more general Corollary 1.5.3 and Theorem 1.6.2.

Recall that q_x is the quadratic approximation of f at x, that is,

$$\begin{aligned}
q_x(y) &:= f(x) + \langle g(x), y - x\rangle + \tfrac{1}{2}\langle y - x, H(x)(y - x)\rangle \\
&= f(x) - \langle n(x), y - x\rangle_x + \tfrac{1}{2}\|y - x\|_x^2,
\end{aligned}$$

where $n(x) := -H(x)^{-1}g(x)$ is the Newton step for f at x.

Theorem 2.2.2. *If* $f \in \mathcal{SC}$, $x \in D_f$ *and* $y \in B_x(x, 1)$, *then*

$$|f(y) - q_x(y)| \leq \frac{\|y - x\|_x^3}{3(1 - \|y - x\|_x)}.$$

Proof. Using Corollary 1.5.3 and Theorem 2.2.1, we have

$$|f(y) - q_x(y)| \leq \|y - x\|_x^2 \int_0^1 \int_0^t \|I - H_x(x + s(y - x))\|_x \, ds \, dt$$

$$\leq \|y - x\|_x^2 \int_0^1 \int_0^t \frac{1}{(1 - s\|y - x\|_x)^2} - 1 \, ds \, dt$$

$$= \|y - x\|_x^3 \int_0^1 \frac{t^2}{1 - t\|y - x\|_x} \, dt$$

$$\leq \frac{\|y - x\|_x^3}{1 - \|y - x\|_x} \int_0^1 t^2 \, dt$$

$$= \frac{\|y - x\|_x^3}{3(1 - \|y - x\|_x)}. \qquad \square$$

Theorem 2.2.3. *Assume* $f \in \mathcal{SC}$ *and* $x \in D_f$. *If* z *minimizes* f *and* $z \in B_x(x, 1)$, *then*

$$x_+ := x - H(x)^{-1} g(x)$$

satisfies

$$\|x_+ - z\|_x \leq \frac{\|x - z\|_x^2}{1 - \|x - z\|_x}.$$

Proof. Using Theorem 1.6.2 and Theorem 2.2.1, simply observe

$$\|x_+ - z\|_x \leq \|x - z\|_x \int_0^1 \|I - H_x(x + t(z - x))\|_x \, dt$$

$$\leq \|x - z\|_x \int_0^1 \frac{1}{(1 - t\|x - z\|_x)^2} - 1 \, dt$$

$$= \frac{\|x - z\|_x^2}{1 - \|x - z\|_x}. \qquad \square$$

The use of the local norm $\|\ \|_x$ in Theorem 2.2.3 to measure the difference $x_+ - z$ makes for a particularly simple proof but does not result in a theorem immediately ready for induction. At x_+, the containment $z \in B_{x_+}(x_+, 1)$ is needed to apply the theorem, i.e., a bound on $\|x_+ - z\|_{x_+}$ rather than a bound on $\|x_+ - z\|_x$. Given that the definition of self-concordance restricts the norms to vary nicely, it is no surprise that the theorem can easily be transformed into a statement ready for induction. For example, substituting into the theorem the inequalities

$$\|x_+ - z\|_z (1 - \|x - z\|_z) \leq \|x_+ - z\|_x$$

2.2. Self-Concordant Functionals

and
$$\|x - z\|_z(1 - \|x - z\|_z) \le \|x - z\|_x \le \frac{\|x - z\|_z}{1 - \|x - z\|_z},$$
as are immediate from the definition of self-concordance when $x \in B_z(z, 1)$, we find as a corollary to the theorem that if $\|x - z\|_z < \frac{1}{2}$, then

$$\|x_+ - z\|_z \le \frac{\|x - z\|_z^2}{(1 - \|x - z\|_z)^2(1 - 2\|x - z\|_z)}. \tag{2.3}$$

Consequently, if one assumes $\|x - z\|_z < \frac{1}{4}$, then

$$\|x_+ - z\|_z < 4\|x - z\|_z^2 \quad (< \tfrac{1}{4}, \text{ so } x_+ \in D_f),$$

and, inductively,
$$\|x_i - z\|_z \le \tfrac{1}{4}(4\|x - z\|_z)^{2^i}, \tag{2.4}$$
where $x_1 = x_+, x_2, \ldots$ is the sequence generated by Newton's method. The bound (2.4) makes apparent the rapid convergence of Newton's method.

Recall $n(x) := -H(x)^{-1}g(x)$, the Newton step. The most elegant proofs of some key results in the ipm theory are obtained by phrasing the analysis in terms of $\|n(x)\|_x$ rather than in terms of $\|z - x\|_x$ or $\|z - x\|_z$. In this regard, the following theorem is especially useful.

Theorem 2.2.4. *Assume $f \in \mathcal{SC}$. If $\|n(x)\|_x < 1$, then*

$$\|n(x_+)\|_{x_+} \le \left(\frac{\|n(x)\|_x}{1 - \|n(x)\|_x}\right)^2.$$

Proof. Assuming $\|n(x)\|_x < 1$, we have

$$\begin{aligned}\|n(x_+)\|_{x_+}^2 &= \|H_x(x_+)^{-1}g_x(x_+)\|_{x_+}^2 \\ &= \langle g_x(x_+), H_x(x_+)^{-1}g_x(x_+)\rangle_x \\ &\le \|H_x(x_+)^{-1}\|_x \|g_x(x_+)\|_x^2.\end{aligned}$$

Since by (2.1) we have
$$\|H_x(x_+)^{-1}\|_x \le \frac{1}{(1 - \|n(x)\|_x)^2},$$
we thus have
$$\|n(x_+)\|_{x_+} \le \frac{\|g_x(x_+)\|_x}{1 - \|n(x)\|_x}.$$

The proof is completed by observing that since $g_x(x) = -n(x)$,

$$\begin{aligned}\|g_x(x_+)\|_x &= \|g_x(x_+) - g_x(x) - n(x)\|_x \\ &= \left\|\int_0^1 [H_x(x + tn(x)) - I]n(x)\,dt\right\|_x \quad \text{(by Proposition 1.5.7)} \\ &\le \|n(x)\|_x \int_0^1 \|I - H_x(x + tn(x))\|_x\,dt\end{aligned}$$

$$\leq \|n(x)\|_x \int_0^1 \frac{1}{(1-t\|n(x)\|_x)^2} - 1 \, dt \quad \text{(by (2.2))}$$
$$= \frac{\|n(x)\|_x^2}{1-\|n(x)\|_x}.$$
□

A rather unsatisfying, but unavoidable, aspect of general convergence results for Newton's method is an assumption of x being sufficiently close to a minimizer z, where "sufficiently close" depends explicitly on z. For general functionals, it is impossible to verify that x is indeed sufficiently close to z without knowing z. For self-concordant functionals, we know that the explicit dependence on z of what constitutes "sufficiently close" can take a particularly simple form (e.g., we know x is sufficiently close to z if $\|x-z\|_z < \frac{1}{4}$), albeit a form which appears still to require knowing z. The next theorem provides means to verify proximity to a minimizer without knowing the minimizer.

Theorem 2.2.5. *Assume $f \in SC$. If $\|n(x)\|_x \leq \frac{1}{4}$ for some $x \in D_f$, then f has a minimizer z and*
$$\|z - x_+\|_x \leq \frac{3\|n(x)\|_x^2}{(1-\|n(x)\|_x)^3}.$$

Thus,
$$\|z - x\|_x \leq \|n(x)\|_x + \frac{3\|n(x)\|_x^2}{(1-\|n(x)\|_x)^3}.$$

Proof. We first prove a weaker result, namely, if $\|n(x)\|_x \leq \frac{1}{9}$, then f has a minimizer z and $\|x - z\|_x \leq 3\|n(x)\|_x$.

Theorem 2.2.2 implies that for all $y \in \bar{B}_x(x, \frac{1}{3})$,
$$|f(y) - q_x(y)| \leq \tfrac{1}{6}\|y - x\|_x^3,$$

and hence, by definition of q_x,
$$f(y) \geq f(x) - \|n(x)\|_x\|y-x\|_x + \tfrac{1}{2}\|y-x\|_x^2 - \tfrac{1}{6}\|y-x\|_x^3$$
$$\geq f(x) - \|n(x)\|_x\|y-x\|_x + \tfrac{1}{3}\|y-x\|_x^2.$$

It follows that if $\|n(x)\|_x \leq \frac{1}{9}$ and $\|y-x\|_x = 3\|n(x)\|_x$, then $f(y) \geq f(x)$. However, it is easily proven that whenever a continuous, convex functional f satisfies $f(y) \geq f(x)$ for all y on the boundary of a compact, convex set S and some x in the interior of S, then f has a minimizer in S. Thus, if $\|n(x)\|_x \leq \frac{1}{9}$, f has a minimizer z and $\|x - z\|_x \leq 3\|n(x)\|_x$.

Now assume $\|n(x)\|_x \leq \frac{1}{4}$. Theorem 2.2.4 implies
$$\|n(x_+)\|_{x_+} \leq \left(\frac{\|n(x)\|_x}{1-\|n(x)\|_x}\right)^2 \leq \tfrac{1}{9}.$$

2.2. Self-Concordant Functionals

Applying the conclusion of the preceding paragraph to x_+ rather than to x, we find that f has a minimizer z and $\|z - x_+\|_{x_+} \leq 3\|n(x_+)\|_{x_+}$. Thus,

$$\|z - x_+\|_x \leq \frac{\|z - x_+\|_{x_+}}{1 - \|n(x)\|_x}$$
$$\leq \frac{3\|n(x_+)\|_{x_+}}{1 - \|n(x)\|_x}$$
$$\leq \frac{3\|n(x)\|_x^2}{(1 - \|n(x)\|_x)^3}.$$

$\square$

2.2.3 Other Properties

We noted that adding a linear functional to a self-concordant functional yields a self-concordant functional. The next two theorems demonstrate other relevant ways of constructing self-concordant functionals. The first theorem shows the set $\mathcal{SC}$ to be closed under addition.

Theorem 2.2.6. *The set $\mathcal{SC}$ is closed under addition; that is, if f_1 and f_2 are self-concordant functionals satisfying $D_{f_1} \cap D_{f_2} \neq \emptyset$, then $f_1 + f_2 : D_{f_1} \cap D_{f_2} \to \mathbb{R}$ is a self-concordant functional.*

Proof. Let $f := f_1 + f_2$. Assume $x \in D_f$. For all v,

$$\langle v, H(x)v \rangle = \langle v, H_1(x)v \rangle + \langle v, H_2(x)v \rangle,$$

that is,

$$\|v\|_x^2 = \|v\|_{x,1}^2 + \|v\|_{x,2}^2.$$

Hence,

$$B_x(x, 1) \subseteq B_{x,1}(x, 1) \cap B_{x,2}(x, 1) \subseteq D_{f_1} \cap D_{f_2} = D_f,$$

as required by the definition of self-concordance.

Note that whenever a, b, c, d are positive numbers,

$$\min\{\tfrac{a}{c}, \tfrac{b}{d}\} \leq \tfrac{a+b}{c+d} \leq \max\{\tfrac{a}{c}, \tfrac{b}{d}\}.$$

Consequently, if $y \in D_f$,

$$\min_{i=1,2} \frac{\|v\|_{y,i}^2}{\|v\|_{x,i}^2} \leq \frac{\|v\|_y^2}{\|v\|_x^2} \leq \max_{i=1,2} \frac{\|v\|_{y,i}^2}{\|v\|_{x,i}^2}.$$

Thus, if $\|y - x\|_x < 1$ (and hence $\|y - x\|_{x,i} < 1$),

$$\frac{\|v\|_y}{\|v\|_x} \leq \max_{i=1,2} \frac{1}{1 - \|y - x\|_{x,i}} \leq \frac{1}{1 - \|y - x\|_x},$$

establishing the upper bound on $\|v\|_y/\|v\|_x$ in the definition of self-concordance. One establishes the lower bound similarly. $\square$

Theorem 2.2.7. *If $f \in SC$, $D_f \subseteq \mathbb{R}^m$, $b \in \mathbb{R}^m$, and $A : \mathbb{R}^n \to \mathbb{R}^m$ is an injective linear operator, then $x \mapsto f(Ax - b)$ is a self-concordant functional, assuming the domain $\{x : Ax - b \in D_f\}$ is nonempty.*

Proof. Denote the functional $x \mapsto f(Ax-b)$ by f'. Let $\| \; \|'_x$ denote the local norm derived from f'. Assuming $x \in D_{f'}$, one easily verifies from the identity $H'(x) = A^*H(Ax - b)A$ that $H'(x)$ is pd and that $\|v\|'_x = \|Av\|_{Ax-b}$ for all v. In particular,

$$AB'_x(x, 1) - b \subseteq B_{Ax-b}(Ax - b, 1) \subseteq D_f$$

and thus

$$B'_x(x, 1) \subseteq \{y : Ay - b \in D_f\} = D_{f'},$$

as required by the definition of self-concordance.

If $\|y - x\|'_x < 1$ (i.e., if $\|A(y - x)\|_{Ax-b} < 1$) and $v \neq 0$, then

$$\frac{\|v\|'_y}{\|v\|'_x} = \frac{\|Av\|_{Ay-b}}{\|Av\|_{Ax-b}}$$

$$\leq \frac{1}{1 - \|A(y - x)\|_{Ax-b}}$$

$$= \frac{1}{1 - \|y - x\|'_x},$$

establishing the upper bound on $\|v\|_y/\|v\|_x$ in the definition of self-concordance. One establishes the lower bound similarly. □

Applying Theorem 2.2.7 with the logarithmic barrier function for the nonnegative orthant in $\mathbb{R}^m$, one verifies self-concordance for the functional

$$x \mapsto -\sum_i \ln(a_i \cdot x - b_i),$$

whose domain consists of the points x satisfying the strict linear inequality constraints $a_i \cdot x > b_i$. This self-concordant functional is important for LPs with constraints written in the form $Ax \geq b$. It, too, is referred to as a "logarithmic barrier function."

To provide the reader with another (logarithmic barrier) functional with which to apply the above theorems, we mention that $x \mapsto -\ln(1 - \|x\|^2)$ is a self-concordant functional with domain the open unit ball. (Verification of self-concordance is made in §2.5.) Given an ellipsoid $\{x : \|Ax\| < r\}$, it then follows from Theorem 2.2.7 that

$$x \mapsto -\ln(r^2 - \|Ax\|^2)$$

is a self-concordant functional whose domain is the ellipsoid, yet another logarithmic barrier function. For an intersection of ellipsoids, one simply adds the functionals for the individual ellipsoids, as justified by Theorem 2.2.6.

Although the values of finite-valued convex functionals with bounded domains (i.e., bounded w.r.t. any reference norm) are always bounded from below, such a functional need not have a minimizer even if it is continuous. This is not the case for self-concordant functionals.

2.2. Self-Concordant Functionals

Theorem 2.2.8. *If $f \in SC$ and the values of f are bounded from below, then f has a minimizer. (In particular, if D_f is bounded, then f has a minimizer.)*

Proof. Choose $\bar{x}$ that satisfies
$$f(\bar{x}) - \tfrac{1}{108} < \inf_x f(x).$$

Letting $y := \bar{x} + \frac{1}{3\|n(\bar{x})\|_{\bar{x}}} n(\bar{x})$, Theorem 2.2.2 shows
$$\begin{aligned} f(y) &\leq f(\bar{x}) - \langle n(\bar{x}), y - \bar{x}\rangle_{\bar{x}} + \tfrac{1}{2}\|y - \bar{x}\|_{\bar{x}}^2 + \frac{\|y - \bar{x}\|_{\bar{x}}^3}{3(1 - \|y - \bar{x}\|_{\bar{x}})} \\ &\leq f(\bar{x}) - \tfrac{1}{3}\|n(\bar{x})\|_{\bar{x}} + \tfrac{1}{2}\left(\tfrac{1}{3}\right)^2 + \tfrac{(1/3)^3}{3(2/3)} \\ &= f(\bar{x}) - \tfrac{1}{3}\|n(\bar{x})\|_{\bar{x}} + \tfrac{2}{27}, \end{aligned}$$

and thus, by choice of $\bar{x}$,
$$\tfrac{1}{3}\|n(\bar{x})\|_{\bar{x}} - \tfrac{2}{27} < \tfrac{1}{108},$$

that is,
$$\|n(\bar{x})\|_{\bar{x}} < 3\left(\tfrac{2}{27} + \tfrac{1}{108}\right) = \tfrac{1}{4}.$$

Theorem 2.2.5 now implies f to have a minimizer. $\square$

The conclusion of the next theorem is trivially verified for important self-concordant functionals like those obtained by adding linear functionals to logarithmic barrier functions. Whereas our definition of self-concordance plays a useful role in simplifying and unifying the analysis of Newton's method for many functionals important to ipm's, it certainly does not simplify the proof of the property established in the next theorem for those same functionals. Nonetheless, for the theory it is important that the property is possessed by all self-concordant functionals.

Theorem 2.2.9. *Assume $f \in SC$ and $\tilde{x} \in \partial D_f$, the boundary of D_f. If the sequence $\{x_i\} \subset D_f$ converges to $\tilde{x}$, then $\lim_i f(x_i) = \infty$ and $\|g(x_i)\| \to \infty$.*

Proof. Assume $x_i \to \tilde{x} \in \partial D_f$. We claim that if $f(x_i) \to \infty$, then $\|g(x_i)\| \to \infty$. Indeed, fix $y \in D_f$. By convexity,
$$f(y) \geq f(x_i) + \langle g(x_i), y - x_i\rangle \geq f(x_i) - \|g(x_i)\|\,\|y - x_i\|.$$

The claim easily follows. Thus it only remains to prove $f(x_i) \to \infty$.

Adding f to the functional $x \mapsto -\ln(\tilde{R} - \|x\|^2)$ where $\tilde{R} > \|\tilde{x}\|^2$, $\|x_i\|^2$ (for all i), one obtains a self-concordant functional $\tilde{f}$ for which $D_{\tilde{f}}$ is bounded and for which $\lim_i \tilde{f}(x_i) = \infty$ iff $\lim_i f(x_i) = \infty$. Consequently, we may assume D_f is bounded.

Assuming D_f is bounded, we will construct from $\{x_i\}$ a sequence $\{y_i\} \subset D_f$ which has a unique limit point, the limit point lying in ∂D_f, and which satisfies
$$\liminf_i f(y_i) \leq \liminf_i f(x_i) - \tfrac{1}{60}.$$

Applying the same construction to the sequence $\{y_i\}$, and so on, we will thus conclude that if $\lim_i f(x_i) \neq \infty$, then f assumes arbitrarily small values, contradicting the lower boundedness of finite-valued convex functionals having bounded domains.

Shortly, we prove $\liminf_i \|n(x_i)\|_{x_i} \geq \frac{1}{5}$. In particular, for sufficiently large i, it follows that $y_i := x_i + \frac{1}{5\|n(x_i)\|_{x_i}} n(x_i)$ is well-defined and, from Theorem 2.2.2,

$$\begin{aligned} f(y_i) &\leq f(x_i) - \langle n(x_i), y_i - x_i \rangle_{x_i} + \tfrac{1}{2}\|y_i - x_i\|_{x_i}^2 + \frac{\|y_i - x_i\|_{x_i}^3}{3(1 - \|y_i - x_i\|_{x_i})} \\ &\leq f(x_i) - \tfrac{1}{25} + \tfrac{1}{50} + \tfrac{1}{300} \\ &= f(x_i) - \tfrac{1}{60}. \end{aligned}$$

Moreover, all limit points of $\{y_i\}$ lie in ∂D_f; indeed, otherwise, passing to a subsequence of $\{y_i\}$ if necessary, there exists $\epsilon > 0$ such that $B(y_i, \epsilon) \subseteq D_f$ for all i, where the ball is w.r.t. a reference norm. However, since y_i and $x_i - (y_i - x_i)$ lie in D_f (because $\|y_i - x_i\|_{x_i} = \frac{1}{5} < 1$), it then follows from convexity of D_f that $B(x_i, \epsilon/2) \subseteq D_f$, contradicting $x_i \to \tilde{x} \in \partial D_f$. Consequently, all limit points of $\{y_i\}$ do indeed lie in ∂D_f. Restricting to a subsequence if necessary, we may assume $\{y_i\}$ has a unique limit point, the limit point lying in ∂D_f.

Finally, we show $\liminf_i \|n(x_i)\|_{x_i} \geq \frac{1}{5}$. Since D_f is bounded, Theorem 2.2.8 shows that f has a minimizer z. Since $B_z(z, 1) \subseteq D_f$ and $x_i \to \tilde{x} \in \partial D_f$, we have $\liminf_i \|x_i - z\|_z \geq 1$. Hence, from the definition of self-concordance, $\liminf_i \|x_i - z\|_{x_i} \geq \frac{1}{2}$. Because f has a most one minimizer, Theorem 2.2.5 then implies $\liminf_i \|n(x_i)\|_{x_i} > \frac{1}{5}$, concluding the proof. □

We close this section with a technical proposition to be called upon later.

If $g(x) + v$ is a vector sufficiently near $g(x)$, there exists $x + u$ close to x such that $g(x + u) = g(x) + v$, a consequence of $H(x)$ being pd and hence invertible. It is useful to quantify "near" and "close" when the inner product is the local inner product, that is, when $H_x(x) = I$ and hence $u \approx v$.

Proposition 2.2.10. *Assume $f \in \mathcal{SC}$ and $x \in D_f$. If $\|v\|_x \leq r$ where $r \leq \frac{1}{4}$, there exists $u \in \bar{B}_x(v, \frac{3r^2}{(1-r)^3})$ such that $g_x(x + u) = g_x(x) + v$.*

Proof. Consider the self-concordant functional

$$y \mapsto -\langle g_x(x) + v, y \rangle_x + f(y), \tag{2.5}$$

a functional whose local inner products agree with those of f. Note that a point z' minimizes the functional iff $g_x(z') = g_x(x) + v$. Under the assumption of the proposition, we thus wish to show z' exists and $u := z' - x$ satisfies $u \in \bar{B}_x(v, \frac{3r^2}{(1-r)^3})$.

Since at x the Newton step for the functional (2.5) is v, the assumption $\|v\|_x \leq r \leq \frac{1}{4}$ allows us to apply Theorem 2.2.5, concluding that a minimizer z' does indeed exist and $\|z' - (x + v)\|_x \leq \frac{3r^2}{(1-r)^3}$. □

2.3 Barrier Functionals

2.3.1 Introduction

A functional f is said to be a *(strongly nondegenerate self-concordant) barrier functional* if $f \in \mathcal{SC}$ and

$$\vartheta_f := \sup_{x \in D_f} \|g_x(x)\|_x^2 < \infty.$$

Let $\mathcal{SCB}$ denote the family of functionals thus defined. We typically refer to elements of $\mathcal{SCB}$ as "barrier functionals."

The definition of barrier functionals is phrased in terms of $\|g_x(x)\|_x$ rather than in terms of the identical quantity $\|n(x)\|_x$ because the importance of barrier functionals for ipm's lies not in applying Newton's method to them directly but rather in applying Newton's method to self-concordant functionals built from them. As mentioned before, for an LP

$$\begin{aligned} \min \quad & c \cdot x \\ \text{s.t.} \quad & Ax = b, \\ & x \geq 0, \end{aligned}$$

the most important self-concordant functionals are those of the form

$$\eta c \cdot x + f|_L(x), \tag{2.6}$$

where $\eta \geq 0$ is a fixed constant, f is the logarithmic barrier function for the nonnegative orthant, and $L := \{x : Ax = b\}$.

When Nesterov and Nemirovskii [15] defined barrier functionals, they referred to ϑ_f as "the parameter of the barrier f." Unfortunately, this can be confused with the phrase "barrier parameter" which predates [15] and refers to the constant η in (2.6). Consequently, we prefer to call ϑ_f the *complexity value of* f, especially because it is the quantity that most often represents f in the complexity analysis of ipm's relying on f.

If one restricts a barrier functional f to a subspace L (or a translation of a subspace), one obtains a barrier functional simply because the local norms for $f|_L$ are the restrictions of the local norms for f and

$$\|g|_{L,x}(x)\|_x = \|P_{L,x} g_x(x)\|_x \leq \|g_x(x)\|_x \leq \sqrt{\vartheta_f}.$$

Clearly, $\vartheta_{f|_L} \leq \vartheta_f$.

The primordial barrier functional is the primordial self-concordant functional, that is, the logarithmic barrier function for the nonnegative orthant, $f(x) := -\sum_j \ln x_j$. Relying on the dot product, so that $g(x)$ is the vector with jth entry $-1/x_j$ and $H(x)$ is the diagonal matrix with jth diagonal entry $1/x_j^2$, we have

$$\|g_x(x)\|_x^2 = g(x) \cdot H(x)^{-1} g(x) = n.$$

Thus, $\vartheta_f = n$.

Now let f denote the logarithmic barrier function for the cone of pd matrices in $\mathbb{S}^{n \times n}$, that is, $f(X) := -\ln \det(X)$. Relying on the trace product, we have, for all $X \in \mathbb{S}^{n \times n}_{++}$,

$$\|g_X(X)\|^2_X = g(X) \circ H(X)^{-1} g(X)$$
$$= \text{trace}(X^{-1} X X^{-1} X)$$
$$= \text{trace}(I)$$
$$= n.$$

Thus, $\vartheta_f = n$.

Finally, let f denote the logarithmic barrier function for the unit ball in $\mathbb{R}^n$, that is, $f(x) := -\ln(1 - \|x\|^2)$ (where $\| \ \| := \langle \ , \ \rangle^{1/2}$ for some inner product). It is not difficult to verify that for x in the unit ball,

$$g(x) = \frac{2}{1 - \|x\|^2} x, \qquad H(x)\Delta x = \frac{2}{1 - \|x\|^2} \Delta x + \frac{4\langle x, \Delta x \rangle}{(1 - \|x\|^2)^2} x,$$

and hence

$$H(x)^{-1} g(x) = \frac{1 - \|x\|^2}{1 + \|x\|^2} x.$$

Consequently,

$$\|g_x(x)\|^2_x = \langle g(x), H(x)^{-1} g(x) \rangle = \frac{2\|x\|^2}{1 + \|x\|^2}.$$

It readily follows that $\vartheta_f = 1$, showing the complexity value need not depend on the dimension n.

Nesterov and Nemirovskii [15] proved a most impressive and theoretically important result, namely, that each open, convex set containing no lines is the domain of a (strongly nondegenerate self-concordant) barrier functional. In fact, they proved the existence of a universal constant C with the property that for each n, if the convex set lies in $\mathbb{R}^n$, there exists a barrier functional whose domain is the set and whose complexity value is bounded by Cn. (I have yet to find a relatively transparent proof of this result, and hence a proof is not contained in this book.) Unfortunately, the result is only of theoretical interest. To rely on self-concordant functionals in devising ipm's, one must be able to readily compute their gradients and Hessians. For the self-concordant functionals proven to exist, from a computational viewpoint one cannot say much more than that the gradients and Hessians exist. By contrast, the importance of the various logarithmic barrier functions we have described lies largely in the ease with which their gradients and Hessians can be computed.

In §2.2 we noted that if a linear functional is added to a self-concordant functional, the resulting functional is self-concordant because the Hessians are unchanged; the definition of self-concordance depends on the Hessians alone. By contrast, adding a linear functional to a barrier functional need not result in a barrier functional. For example, consider the univariate barrier functional $x \mapsto -\ln x$ and the functional $x \mapsto x - \ln x$.

The set $\mathcal{SCB}$, like $\mathcal{SC}$, is closed under addition.

Theorem 2.3.1. *If $f_1, f_2 \in \mathcal{SCB}$ and $D_{f_1} \cap D_{f_2} \neq \emptyset$, then $f := f_1 + f_2 \in \mathcal{SCB}$ (where $D_f = D_{f_1} \cap D_{f_2}$) and $\vartheta_f \leq \vartheta_{f_1} + \vartheta_{f_2}$.*

2.3. Barrier Functionals

Proof. Assume $x \in D_f$. Let the reference inner product $\langle \, , \, \rangle$ be the local inner product at x defined by f. Thus, $I = H(x) = H_1(x) + H_2(x)$. In particular, $H_1(x)$ and $H_2(x)$ commute, i.e., $H_1(x)H_2(x) = H_2(x)H_1(x)$. Consequently, so do $H_1(x)^{1/2}$ and $H_2(x)^{1/2}$.

For brevity, let $H_i := H_i(x)$ and $g_i := g_i(x)$ for $i = 1, 2$.

To prove the inequality in the statement of the theorem, it suffices to show
$$\|g_1 + g_2\|^2 \leq \langle g_1, H_1^{-1} g_1 \rangle + \langle g_2, H_2^{-1} g_2 \rangle,$$
since, by definition, the quantity on the right is bounded from above by $\vartheta_{f_1} + \vartheta_{f_2}$.

Defining $v_i := H_i^{-1/2} g_i$ for $i = 1, 2$, we have
$$\begin{aligned}
\|g_1 + g_2\|^2 &= \|g_1\|^2 + 2\langle g_1, g_2 \rangle + \|g_2\|^2 \\
&= \langle v_1, H_1 v_1 \rangle + 2\langle H_1^{1/2} v_1, H_2^{1/2} v_2 \rangle + \langle v_2, H_2 v_2 \rangle \\
&= \langle v_1, (I - H_2) v_1 \rangle + 2\langle H_1^{1/2} v_1, H_2^{1/2} v_2 \rangle + \langle v_2, (I - H_1) v_2 \rangle \\
&= \langle v_1, v_1 \rangle + \langle v_2, v_2 \rangle - \|H_2^{1/2} v_1 - H_1^{1/2} v_2\|^2 \\
&\leq \langle v_1, v_1 \rangle + \langle v_2, v_2 \rangle \\
&= \langle g_1, H_1^{-1} g_1 \rangle + \langle g_2, H_2^{-1} g_2 \rangle,
\end{aligned}$$
where the fourth equality relies on $H_1^{1/2}$ and $H_2^{1/2}$ commuting. □

The set $\mathcal{SCB}$, like $\mathcal{SC}$, is closed under composition with injective linear maps.

Theorem 2.3.2. *If $f \in \mathcal{SCB}$, $D_f \subseteq \mathbb{R}^m$, $b \in \mathbb{R}^m$, and $A : \mathbb{R}^n \to \mathbb{R}^m$ is an injective linear operator, then $x \mapsto f(Ax - b)$ is a barrier functional—assuming the domain $\{x : Ax - b \in D_f\}$ is nonempty—and its complexity value does not exceed ϑ_f.*

Proof. Assume $Ax - b \in D_f$. Endow $\mathbb{R}^n$ with an arbitrary reference inner product and let the reference inner product on $\mathbb{R}^m$ be the local inner product for f at $Ax - b$. Denoting the functional $x \mapsto f(Ax - b)$ by f', we then have $g'(x) = A^* g(Ax - b)$, $H'(x) = A^* H(Ax - b) A = A^* A$, and $\|g(Ax - b)\|^2 \leq \vartheta_f$. Thus,
$$\begin{aligned}
\langle g'(x), H'(x)^{-1} g'(x) \rangle &= \langle g(Ax - b), A(A^*A)^{-1} A^* g(Ax - b) \rangle \\
&\leq \|A(A^*A)^{-1} A^*\| \, \|g(Ax - b)\|^2 \\
&\leq \vartheta_f.
\end{aligned}$$

(The last inequality is due to the operator being an orthogonal projection operator; the operator has norm equal to one.) □

With regards to theory, the following theorem is perhaps the most useful tool in establishing properties possessed by all barrier functionals. The inner product is arbitrary.

Theorem 2.3.3. *Assume $f \in \mathcal{SCB}$. If $x, y \in D_f$, then*
$$\langle g(x), y - x \rangle < \vartheta_f.$$

Proof. We wish to prove $\phi'(0) < \vartheta_f$, where ϕ is the univariate functional defined by $\phi(t) := f(x + t(y - x))$. In doing so, we may assume $\phi'(0) > 0$ and hence, by convexity of ϕ, $\phi'(t) > 0$ for all $t \geq 0$ in the domain of ϕ.

Let $v(t) := x + t(y - x)$. Assuming $t \geq 0$ is in the domain of ϕ,
$$\frac{\phi''(t)}{\phi'(t)^2} = \frac{\langle y - x, y - x \rangle_{v(t)}}{\langle g_{v(t)}(v(t)), y - x \rangle_{v(t)}^2}$$
$$\geq \frac{1}{\|g_{v(t)}(v(t))\|_{v(t)}^2}$$
$$\geq \frac{1}{\vartheta_f},$$
and hence, for all $s \geq 0$ in the domain of ϕ,
$$\int_0^s \frac{\phi''(t)}{\phi'(t)^2} \, dt \geq \frac{s}{\vartheta_f}.$$
Thus,
$$\left. \frac{-1}{\phi'(t)} \right|_0^s \geq \frac{s}{\vartheta_f},$$
that is,
$$\phi'(s) \geq \frac{\vartheta_f \phi'(0)}{\vartheta_f - s\phi'(0)}.$$
Consequently, the domain of the convex functional ϕ is contained in the open interval $(-\infty, \vartheta_f/\phi'(0))$. In particular, $\vartheta_f/\phi'(0)$ is not in the domain. Since $s = 1$ is in the domain of ϕ we thus have $1 < \vartheta_f/\phi'(0)$. □

2.3.2 Analytic Centers

The next theorem implies that for each x in the domain of a barrier functional, the ball $B_x(x, 1)$ is, to within a factor of $4\vartheta_f + 1$, the largest among all ellipsoids centered at x which are contained in the domain. A consequence is that no (strongly nondegenerate self-concordant) barrier functional has a domain containing a line.

Theorem 2.3.4. *Assume* $f \in \mathcal{SCB}$. *If* $x, y \in D_f$ *satisfy* $\langle g(x), y - x \rangle \geq 0$, *then* $y \in B_x(x, 4\vartheta_f + 1)$.

Proof. Restricting f to the line through x and y, we may assume f is univariate. Viewing the line as $\mathbb{R}$ with values increasing as one travels from x to y, the assumption $\langle g(x), y - x \rangle \geq 0$ is then equivalent to $g(x) \geq 0$, i.e., $g(x)$ is a nonnegative number.

Let v denote the smallest nonnegative number for which $\|g_x(x) + v\|_x \geq \frac{1}{4}$. Since $g_x(x) \geq 0$, we have $\|v\|_x \leq \frac{1}{4}$. Applying Proposition 2.2.10, we find there exists u satisfying
$$u \in \bar{B}_x\left(v, \tfrac{4}{9}\right) \quad \text{and} \quad \|g_x(x+u)\|_x = \|g_x(x) + v\|_x \geq \tfrac{1}{4}.$$
Note that $\|u\|_x < 1$.

Theorem 2.3.3 implies
$$\vartheta_f \geq \langle g_x(x+u), y - (x+u) \rangle_x$$
$$= \langle g_x(x) + v, y - x \rangle_x - \langle g_x(x) + v, u \rangle_x$$
$$\geq \tfrac{1}{4} \|y - x\|_x - \langle g_x(x) + v, u \rangle_x,$$

2.3. Barrier Functionals

where the last inequality makes use of $g_x(x) + v$ and $y - x$ both being nonnegative. However, since $\|g_x(x) + v\|_x > \frac{1}{4}$ only if $v = 0$ (and hence only if $u = 0$), we have

$$\langle g_x(x) + v, u \rangle_x \leq \tfrac{1}{4} \|u\|_x < \tfrac{1}{4}.$$

Thus,
$$\vartheta_f > \tfrac{1}{4} \|y - x\|_x - \tfrac{1}{4},$$

from which the theorem is immediate. □

Minimizers of barrier functionals are called *analytic centers*. The following corollary gives meaning to the term "center."

Corollary 2.3.5. *Assume $f \in \mathcal{SCB}$. If z is the analytic center for f, then*

$$B_z(z, 1) \subseteq D_f \subseteq B_z(z, 4\vartheta_f + 1).$$

Proof. Since $f \in \mathcal{SC}$, the leftmost containment is by assumption. The rightmost containment is immediate from Theorem 2.3.4 since $g(z) = 0$. □

Corollary 2.3.5 suggests that if one was to choose a single inner product as being especially natural for a barrier functional with bounded domain, the local inner product at the analytic center would be an appropriate choice because the resulting balls conform to the shape of the domain.

When do analytic centers exist? The answer is given by the following corollary.

Corollary 2.3.6. *If $f \in \mathcal{SCB}$, then f has an analytic center iff D_f is bounded.*

Proof. The proof is immediate from Theorem 2.2.8 and Corollary 2.3.5. □

2.3.3 Optimal Barrier Functionals

In the complexity analysis of ipm's, it is desirable to have barrier functionals with small complexity values. However, there is a positive threshold below which the complexity values of no barrier functionals fall. Nesterov and Nemirovskii [15] prove that $\vartheta_f \geq 1$ for all $f \in \mathcal{SCB}$. To understand why there is indeed a lower bound, assume $\vartheta_f \leq \frac{1}{16}$ for some $f \in \mathcal{SCB}$. Since $g_x(x) = -n(x)$, Theorem 2.2.5 then implies f has a (unique) minimizer z and all $x \in D_f$ satisfy $\|z - x\|_x \leq \frac{25}{36}$. However, by choosing x so that in the line L through x and z, the distance from x to the boundary of $D_f \cap L$ is smaller than the distance from x to z, the containment $B_x(x, 1) \subseteq D_f$ implies $\|z - x\|_x \geq 1$, a contradiction. Hence $\vartheta_f > \frac{1}{16}$ for all $f \in \mathcal{SCB}$.

Likewise, by Theorem 2.2.5, if $f \in \mathcal{SCB}$ and D_f is unbounded—hence f has no minimizer—then $\|g_x(x)\|_x > \ell := \frac{1}{4}$ for *all* $x \in D_f$.

It is worth noting that any universal lower bound ℓ as in the preceding paragraph implies a lower bound $n\ell \leq \vartheta_f$ for each barrier functional f whose domain is the nonnegative orthant $\mathbb{R}^n_{++}$. Indeed, let e denote the vector of all ones and let e_j denote the jth unit vector. Consider the univariate barrier functional f_j obtained by restricting f to the line through e in the direction e_j. Let g_e denote the gradient of f w.r.t. $\langle \, , \, \rangle_e$ and let $g_{j,e}$ denote the gradient of f_j w.r.t. the restricted inner product. Since D_{f_j} is unbounded, and hence f_j does not have

an analytic center, it is readily proven (without making use of a particular inner product) that $\langle g_{j,e}(e), e_j \rangle_e \leq 0$. Since $g_{j,e}$ and e_j are colinear (because D_{f_j} is one-dimensional), it follows that
$$\langle g_{j,e}(e), e_j \rangle_e = -\|g_{j,e}(e)\|_e \|e_j\|_e.$$
Noting that $\|e_j\|_e \geq 1$ because $e - e_j \notin D_f$, we thus have
$$\langle g_{j,e}(e), e_j \rangle_e \leq -\|g_{j,e}(e)\|_e \leq -\ell.$$
Hence,
$$\begin{aligned} n\ell &\leq \sum_j \langle g_{j,e}(e), -e_j \rangle_e \\ &= \sum_j \langle g_e(e), -e_j \rangle_e \\ &= \langle g_e(e), 0 - e \rangle_e \\ &\leq \vartheta_f, \end{aligned}$$
the last inequality by Theorem 2.3.3.

In light of the preceding paragraphs, we see that with regards to the complexity value, the logarithmic barrier function for the nonnegative orthant $\mathbb{R}^n_{++}$ is the optimal barrier functional having domain $\mathbb{R}^n_{++}$. Likewise, viewing $\mathbb{R}^n$ as a subspace of $\mathbb{S}^{n \times n}$, we can conclude that the logarithmic barrier function for the cone of pd matrices is the optimal barrier functional having that cone as its domain.

2.3.4 Other Properties

For arbitrary inner products, the bounds $\|g_x(x)\|_x \leq \sqrt{\vartheta_f}$ imply nothing about the quantities $\|g(x)\|$. However, the bounds do imply bounds on the quantities $\|g_y(x)\|_y$ for all $y \in D_f$. This is the subject of the next proposition. First, a definition.

For x in an arbitrary bounded convex set D, a natural way of measuring the relative nearness of x to the boundary of D, in a manner that is independent of a particular norm, is the quantity known as *the symmetry of D about x*, denoted $\mathrm{sym}(x, D)$. This quantity is defined in terms of the set $\mathcal{L}(x, D)$ consisting of all lines through x which intersect D in an interval of positive length. (If D is lower-dimensional, most lines through x will not be in $\mathcal{L}(x, D)$.) If x is an endpoint of $L \cap D$ for some $L \in \mathcal{L}(x, D)$, define $\mathrm{sym}(x, D) := 0$. Otherwise, for each $L \in \mathcal{L}(x, D)$, letting $r(L)$ denote the ratio of the length of the smaller to the larger of the two intervals in $L \cap (D \setminus \{x\})$, define
$$\mathrm{sym}(x, D) := \inf_{L \in \mathcal{L}(x, D)} r(L).$$
Clearly, if D is an ellipsoid centered at x, then $\mathrm{sym}(x, D) = 1$, "perfect symmetry." Corollary 2.3.5 implies that if z is the analytic center for a barrier functional f, then $\mathrm{sym}(z, D_f) \geq 1/(4\vartheta_f + 1)$.

Proposition 2.3.7. *Assume $f \in \mathcal{SCB}$. For all $x, y \in D_f$,*
$$\|g_y(x)\|_y \leq \left(1 + \frac{1}{\mathrm{sym}(x, D_f)}\right) \vartheta_f.$$

2.3. Barrier Functionals

Proof. For brevity, let $s := \text{sym}(x, D_f)$. Assuming $x, y \in D_f$, note that $w := y + (1 + s)(x - y) \in \bar{D}_f$ (the closure of D_f) since w, x, y are colinear and $\|w - x\| = s\|y - x\|$. Since $B_y(y, 1) \subseteq D_f$ and D_f is convex, we thus have

$$\tfrac{1}{1+s} w + \tfrac{s}{1+s} B_y(y, 1) \subseteq D_f,$$

that is, $B_y(x, \tfrac{s}{1+s}) \subseteq D_f$. Consequently,

$$\begin{aligned}
\|g_y(x)\|_y &= \max_{v \in B_y(x,1)} \langle g_y(x), v - x \rangle_y \\
&= \max_{v \in B_y(x, \tfrac{s}{1+s})} \tfrac{1+s}{s} \langle g_y(x), v - x \rangle_y \\
&\leq \max_{v \in D_f} \tfrac{1+s}{s} \langle g_y(x), v - x \rangle_y \\
&\leq \tfrac{1+s}{s} \vartheta_f,
\end{aligned}$$

the last inequality by Theorem 2.3.3. □

Theorem 2.2.9 shows that $f(x_i) \to \infty$ if f is a self-concordant functional and $\{x_i\}$ converges to a point in the boundary of D_f. To close this subsection, we present a theorem that implies the rate at which $f(x_i)$ goes to ∞ is "slow" if f is a barrier functional.

Theorem 2.3.8. *Assume $f \in SCB$ and $x \in D_f$. If $y \in \bar{D}_f$, then for all $0 < t \leq 1$,*

$$f(y + t(x - y)) \leq f(x) - \vartheta_f \ln t.$$

Proof. For $s \geq 0$ let $x(s) := y + e^{-s}(x - y)$ and consider the univariate functional $\phi(s) := f(x(s))$. Relying on the chain rule, observe

$$\begin{aligned}
\phi(s) &= \phi(0) + \int_0^s \phi'(t) \, dt \\
&= f(x) + \int_0^s \langle g(x(t)), -e^{-t}(x - y) \rangle \, dt \\
&= f(x) + \int_0^s \langle g(x(t)), y - x(t) \rangle \, dt \\
&\leq f(x) + \int_0^s \vartheta_f \, dt \\
&= f(x) + s\vartheta_f,
\end{aligned}$$

the inequality due to Theorem 2.3.3. Hence, for $0 < t \leq 1$,

$$f(y + t(x - y)) = \phi(-\ln t) \leq f(x) - \vartheta_f \ln(t). \qquad \square$$

2.3.5 Logarithmic Homogeneity

In Chapter 3, when we tie ipm's to duality theory, attention will often focus on a particular type of barrier functional f whose domain is the interior K° of a closed, convex cone K (if

$x_1, x_2 \in K$ and $t_1, t_2 \geq 0$, then $t_1 x_1 + t_2 x_2 \in K$). A barrier functional $f : K^\circ \to \mathbb{R}$ is said to be *logarithmically homogeneous* if for all $x \in K^\circ$ and $t > 0$,

$$f(tx) = f(x) - \vartheta_f \ln t. \tag{2.7}$$

It is easily established that the logarithmic barrier functions for the nonnegative orthant and the cone of psd matrices are logarithmically homogeneous, as are barrier functionals of the form $x \mapsto f(Ax)$ where f is logarithmically homogeneous. Another important example of a logarithmically homogeneous barrier functional is

$$f(x) := -\ln\left(x_n^2 - \sum_{j=1}^{n-1} x_j^2\right),$$

the domain of this functional being the interior of the *second-order cone*

$$K := \left\{ x \in \mathbb{R}^n : \sum_{j=1}^{n-1} x_j^2 \leq x_n^2 \text{ and } x_n \geq 0 \right\}.$$

It has complexity value $\vartheta_f = 2$. As with the standard barrier functionals for the nonnegative orthant and the cone of psd matrices, it is referred to as a logarithmic barrier function.

The following theorem provides a characterization of logarithmic homogeneity as well as other properties useful in analysis.

Theorem 2.3.9. *A self-concordant functional $f : K^\circ \to \mathbb{R}$ is a logarithmically homogeneous barrier functional iff for all $x \in K^\circ$ and $t > 0$*

$$g(tx) = \tfrac{1}{t} g(x). \tag{2.8}$$

Moreover, a logarithmically homogeneous barrier functional satisfies the identities

$$H(tx) = \tfrac{1}{t^2} H(x), \quad g_x(x) = -x, \quad \text{and} \quad \|g_x(x)\|_x = \sqrt{\vartheta_f}.$$

Proof. Assume f is a logarithmically homogeneous barrier functional. To prove (2.8), simply differentiate both sides of (2.7) w.r.t. x. Similarly, differentiating both sides of (2.8) w.r.t. x gives $H(tx) = \tfrac{1}{t^2} H(x)$.

Now assume f is a self-concordant functional satisfying (2.8) for all $x \in K^\circ$ and $t > 0$. Differentiating both sides of (2.8) w.r.t. t gives

$$H(tx)x = -\tfrac{1}{t^2} g(x).$$

For $t = 1$ this is the same as $g_x(x) = -x$.

Since $g_x(x) = -x$, we have

$$\|x\|_x^2 = \|g_x(x)\|_x^2 = -\langle x, g(x) \rangle. \tag{2.9}$$

The gradient of the rightmost quantity in (2.9) as a functional of x is

$$-g(x) - H(x)x = -H(x)(g_x(x) + x) = 0.$$

2.4. Primal Algorithms

Hence, the three quantities in (2.9) are independent of x. In particular, the middle quantity is bounded above as a functional of x, and thus f is not only self-concordant, it is also a barrier functional. Moreover, the independence from x of the three quantities implies each of them is equal to ϑ_f.

Finally,
$$\begin{aligned} f(tx) &= f(x) + \int_1^t \langle g(sx), x \rangle \, ds \\ &= f(x) + \int_1^t \tfrac{1}{s} \langle g(x), x \rangle \, ds \\ &= f(x) - \vartheta_f \int_1^t \tfrac{1}{s} \, ds \\ &= f(x) - \vartheta_f \ln t, \end{aligned}$$

showing f is logarithmically homogeneous. □

Using the "$g(tx) = \tfrac{1}{t} g(x)$" characterization of logarithmic homogeneity provided by the theorem, it is simple to prove that a sum of logarithmically homogeneous barrier functionals is itself a logarithmically homogeneous barrier functional.

In closing our discussion of logarithmic homogeneity, we recall that in §2.3.3 it was established that for all barrier functionals with unbounded domains, $\vartheta_f \geq \tfrac{1}{4}$. We noted that with greater effort, Nesterov and Nemirovskii proved $\vartheta_f \geq 1$. The stronger inequality is easily established for logarithmically homogeneous barrier functionals. Indeed, since f is self-concordant and 0 lies on the boundary of $K = \bar{D}_f$, we must have $\|0 - x\|_x \geq 1$ for all $x \in K^\circ$, that is, $\|g_x(x)\|_x \geq 1$ for all $x \in D_f$.

2.4 Primal Algorithms

2.4.1 Introduction

The importance of a barrier functional f lies not in itself but in that it can be used to efficiently solve optimization problems of the form

$$\begin{aligned} \min \quad & \langle c, x \rangle \\ \text{s.t.} \quad & x \in \bar{D}_f, \end{aligned} \qquad (2.10)$$

where $\bar{D}_f$ denotes the closure of D_f. Among many other problems, linear programs are of this form. Specifically, restricting the logarithmic barrier function for the nonnegative orthant to the affine space $\{x : Ax = b\}$, we obtain a barrier functional f for which

$$D_f = \{x : Ax = b, \ x > 0\}.$$

Similarly for SDP.

Let val denote the optimal value of the optimization problem (2.10).

Path-following ipm's solve (2.10) by following the *central path* (a.k.a. "the path of analytic centers" and "the central trajectory"), the path consisting of the minimizers $z(\eta)$ of the self-concordant functionals

$$f_\eta(x) := \eta \langle c, x \rangle + f(x)$$

for $\eta > 0$. It is readily proven when D_f is bounded that the central path begins at the analytic center z of f and consists of the minimizers of the barrier functionals $f|_{L(v)}$ obtained by restricting f to the affine spaces

$$L(v) := \{x : \langle c, x \rangle = v\}$$

for v satisfying val $< v < \langle c, z \rangle$. Similarly, when D_f is unbounded, the central path consists of the minimizers of the barrier functionals $f|_{L(v)}$ for v satisfying val $< v$.

In the literature, it is standard to define the central path, and to do ipm analysis, with the functionals

$$x \mapsto \langle c, x \rangle + \mu f(x), \tag{2.11}$$

where $\mu > 0$, rather than with the functionals f_η. The difference is only cosmetic. The minimizer $z(\eta)$ of f_η is the minimizer of the functional (2.11) for $\mu = 1/\eta$. Similarly, Newton's method applied to minimizing f_η produces exactly the same sequence of points as Newton's method applied to minimizing the functional (2.11) for $\mu = 1/\eta$. Whether one relies on the functionals f_η or the functionals (2.11), one arrives at exactly the same ipm results, the only difference being insignificant changes in minor algebraic steps along the way. The reason we use the functionals f_η rather than the customary functionals (2.11) is that the local inner products for the functionals f_η are identical to those for f, whereas the local inner products for the functionals (2.11) depend on μ. Since a goal of this book is to elucidate the geometry underlying ipm's, it is more natural for us to rely on the functionals f_η.

We observe that for each $y \in D_f$, the optimization problem (2.10) is equivalent to

$$\begin{aligned} \min \quad & \langle c_y, x \rangle_y \\ \text{s.t.} \quad & x \in \bar{D}_f, \end{aligned}$$

where $c_y := H(y)^{-1} c$. (In other words, the objective vector is c_y w.r.t. $\langle\, ,\, \rangle_y$.)

The desirability of following the central path is made evident by considering the objective values $\langle c, z(\eta) \rangle$. Since $g(z(\eta)) = -\eta c$, Theorem 2.3.3 implies for all $y \in D_f$,

$$\begin{aligned} \langle c, z(\eta) \rangle - \langle c, y \rangle &= \tfrac{1}{\eta} \langle g(z(\eta)), y - z(\eta) \rangle \\ &< \tfrac{1}{\eta} \vartheta_f, \end{aligned}$$

and hence

$$\langle c, z(\eta) \rangle \leq \text{val} + \tfrac{1}{\eta} \vartheta_f. \tag{2.12}$$

Moreover, the point $z(\eta)$ is well centered in the sense that all feasible points y with objective value at most $\langle c, z(\eta) \rangle$ satisfy $y \in B_{z(\eta)}(z(\eta), 4\vartheta_f + 1)$, a consequence of Theorem 2.3.4 and $g(z(\eta)) = -\eta c$.

Path-following ipm's follow the central path approximately, generating points near the central path where "near" is measured by local norms. If a point y is computed for which $\|y - z(\eta)\|_{z(\eta)}$ is small, then, relatively, the objective value at y will not be much worse than at $z(\eta)$, and hence (2.12) implies a bound on $\langle c, y \rangle$. In fact, if x is an *arbitrary* point in D_f and y is a point for which $\|y - x\|_x$ is small, then, relatively, the objective value at y will not be much worse than at x. To make this precise, first observe $B_x(x, 1) \subseteq D_f$ implies

2.4. Primal Algorithms

$x - tc_x \in D_f$ if $0 \leq t < 1/\|c_x\|_x$. Since the objective value at $x - tc_x$ is $\langle c, x \rangle - t\|c_x\|_x^2$, we thus have
$$\|c_x\|_x \leq \langle c, x \rangle - \text{val}. \tag{2.13}$$
Hence for all $y \in \mathbb{R}^n$,
$$\begin{aligned}\frac{\langle c, y \rangle - \text{val}}{\langle c, x \rangle - \text{val}} &= 1 + \frac{\langle c_x, y - x \rangle_x}{\langle c, x \rangle - \text{val}} \\ &\leq 1 + \frac{\|c_x\|_x \|y - x\|_x}{\langle c, x \rangle - \text{val}} \\ &\leq 1 + \|y - x\|_x.\end{aligned}$$

In particular, using (2.12),
$$\langle c, y \rangle \leq \text{val} + \tfrac{1}{\eta}\vartheta_f(1 + \|y - z(\eta)\|_{z(\eta)}). \tag{2.14}$$

Before discussing algorithms, we record a piece of notation: Let $n_\eta(x)$ denote the Newton step for f_η at x, that is,
$$\begin{aligned}n_\eta(x) &:= -H(x)^{-1}(\eta c + g(x)) \\ &= -(\eta c_x + g_x(x)).\end{aligned}$$

2.4.2 The Barrier Method

"Short-step" ipm's follow the central path most closely, generating sequences of points which are *all* near the path. We now present and analyze an elementary short-step ipm, the "barrier method."

Assume, initially, we know $\eta_1 > 0$ and x_1 such that x_1 is "near" $z(\eta_1)$, that is, x_1 is near the minimizer for the functional f_{η_1}. In the algorithm, we increase η_1 to a value η_2 and then apply Newton's method to approximate $z(\eta_2)$, thus obtaining a point x_2. Assuming only one iteration of Newton's method is applied,
$$x_2 := x_1 + n_{\eta_2}(x_1).$$

Continuing this procedure indefinitely (i.e., increasing η, applying Newton's method, increasing η, etc.), we have the barrier method.

One would like η_2 to be much larger than η_1. However, if η_2 is "too" large relative to η_1, Newton's method can fail to approximate $z(\eta_2)$; in fact, it can happen that $x_2 \notin D_f$, bringing the algorithm to a halt. The main goal in analyzing the barrier method is to prove that η_2 can be larger than η_1 by a reasonable amount without the algorithm losing sight of the central path.

In analyzing the barrier method, it is most natural to rely on the length of Newton steps to measure proximity to the central path. We will assume x_1 is near $z(\eta_1)$ in the sense that $\|n_{\eta_1}(x_1)\|_{x_1}$ is small. Keep in mind that the Newton step taken by the algorithm is $n_{\eta_2}(x_1)$, not $n_{\eta_1}(x_1)$. The relevance of $n_{\eta_1}(x_1)$ for $n_{\eta_2}(x_1)$ is due to the following easily proven relation:
$$n_{\eta_2}(x) = \tfrac{\eta_2}{\eta_1} n_{\eta_1}(x) + \left(\tfrac{\eta_2}{\eta_1} - 1\right) g_x(x).$$

In particular,
$$\|n_{\eta_2}(x)\|_x \le \tfrac{\eta_2}{\eta_1}\|n_{\eta_1}(x)\|_x + |\tfrac{\eta_2}{\eta_1} - 1|\sqrt{\vartheta_f}. \tag{2.15}$$

Besides the bound (2.15), a crucial ingredient in the analysis is a bound on $\|n_{\eta_2}(x_2)\|_{x_2}$ in terms of $\|n_{\eta_2}(x_1)\|_{x_1}$. Theorem 2.2.4 provides an appropriate bound: if $\|n_{\eta_2}(x_1)\|_{x_1} < 1$, then

$$\|n_{\eta_2}(x_2)\|_{x_2} \le \left(\frac{\|n_{\eta_2}(x_1)\|_{x_1}}{1 - \|n_{\eta_2}(x_1)\|_{x_1}}\right)^2. \tag{2.16}$$

Suppose we determine values $\alpha > 0$ and $\beta > 1$ such that if we define

$$\gamma := \alpha\beta + (\beta - 1)\sqrt{\vartheta_f},$$

then $\gamma < 1$ and

$$\left(\frac{\gamma}{1 - \gamma}\right)^2 \le \alpha.$$

By requiring $\|n_{\eta_1}(x_1)\|_{x_1} \le \alpha$ and $1 \le \tfrac{\eta_2}{\eta_1} \le \beta$, we then find from (2.15) that

$$\|n_{\eta_2}(x_1)\|_{x_1} \le \gamma,$$

and thus, from (2.16),

$$\|n_{\eta_2}(x_2)\|_{x_2} \le \alpha.$$

Consequently, x_2 will be close to the central path like x_1. Continuing, by requiring $1 \le \tfrac{\eta_3}{\eta_2} \le \beta$, x_3 will be close to the central path, too, and so on. Hence, we will have determined a value β such that if one has an initial point appropriately close to the central path, and if one never increases the barrier parameter from η to more than $\beta\eta$, the barrier method will follow the central path, always generating points close to it.

The reader can verify, for example, that

$$\alpha = \frac{1}{9} \quad \text{and} \quad \beta := 1 + \frac{1}{8\max\{1, \sqrt{\vartheta_f}\}} \quad \left(= 1 + \frac{1}{8\sqrt{\vartheta_f}}\right)$$

satisfy the relations. Now we have a "safe" value for β. Relying on it, the algorithm is guaranteed to stay on track. It is a remarkable aspect of ipm's that safe values for quantities like β depend only on the complexity value ϑ_f of the underlying barrier functional f. Concerning LPs,

$$\begin{aligned} \min \quad & c \cdot x \\ \text{s.t.} \quad & Ax = b, \\ & x \ge 0, \end{aligned}$$

if one relies on the logarithmic barrier function for the strictly nonnegative orthant $\mathbb{R}^n_{++}$, then $\gamma = (1 + \tfrac{1}{8\sqrt{n}})$ is safe *regardless of A, b, and c*.

Assuming at each iteration of the barrier method the parameter η is never increased by more than $1 + 1/8\sqrt{\vartheta_f}$, we now know that for each x generated by the algorithm, there corresponds $z(\eta)$ which x approximates in that $\|n_\eta(x)\|_x \le \tfrac{1}{9}$; hence, by Theorem 2.2.5, $\|x - z(\eta)\|_x \le \tfrac{1}{6}$; thus, by the definition of self-concordance, $\|x - z(\eta)\|_{z(\eta)} \le \tfrac{1}{5}$. All points generated by the algorithm lie within distance $\tfrac{1}{5}$ of the central path.

2.4. Primal Algorithms

Assuming that at each iteration of the barrier method, the parameter η is increased by exactly the factor $1 + 1/8\sqrt{\vartheta_f}$, the number of iterations required to increase the parameter from an initial value η_1 to some value $\eta > \eta_1$ is

$$\begin{aligned} i &= \frac{\ln(\eta/\eta_1)}{\ln(1 + 1/8\sqrt{\vartheta_f})} \\ &\leq 10\sqrt{\vartheta_f}\ln(\eta/\eta_1), \\ &= O(\sqrt{\vartheta_f}\log(\eta/\eta_1)), \end{aligned} \quad (2.17)$$

where in the inequality we rely on $\vartheta_f \geq 1$ (as discussed in §2.3.3). Hence, from (2.14), given $\epsilon > 0$,

$$O\left(\sqrt{\vartheta_f}\log\left(\frac{\vartheta_f}{\epsilon\eta_1}\right)\right) \quad (2.18)$$

iterations suffice to produce x satisfying $\langle c, x \rangle \leq \mathrm{val} + \epsilon$.

We have been assuming that an initial point x_1 near the central path is available. What if, instead, we know only some arbitrary point $x' \in D_f$? How might we use the barrier method to solve the optimization problem efficiently? We now describe a simple approach, assuming D_f is bounded and hence f has an analytic center.

Consider the optimization problem obtained by replacing the objective vector c with $-g(x')$. The central path then consists of the minimizers $z'(\nu)$ of the self-concordant functionals

$$f'_\nu(x) := -\nu \langle g(x'), x \rangle + f(x).$$

The point x' is on the central path for this optimization problem. In fact, $x' = z'(\nu)$ for $\nu = 1$.

Let $n'_\nu(x)$ denote the Newton step for f'_ν at x.

Rather than increasing the parameter ν, we decrease it toward zero, following the central path to the analytic center z of f. From there, we switch to following the central path $\{z(\eta) : \eta > 0\}$ as before.

We showed η can safely be increased by a factor of $1 + 1/8\sqrt{\vartheta_f}$. Brief consideration of the analysis shows it is also safe to decrease η by a factor $1 - 1/8\sqrt{\vartheta_f}$ and, hence, safe to decrease ν by that factor. Thus, to complete our understanding of the difficulty of following the path $\{z'(\nu) : \nu > 0\}$, and then the path $\{z(\eta) : \eta > 0\}$, it remains only to understand the process of switching paths.

One way to know when it is safe to switch paths is to compute the length of the gradients for f at the points x generated in following the path $\{z'(\nu) : \nu > 0\}$. Once one encounters a point x for which, say, $\|g_x(x)\|_x^* \leq \frac{1}{6}$, one can safely switch paths. For then, by choosing $\eta_1 = 1/12 \|c_x\|_x^*$, we find the Newton step for f_{η_1} at x satisfies

$$\|n_{\eta_1}(x)\|_x = \|\eta_1 c_x + g_x(x)\|_x^* \leq \tfrac{1}{12} + \tfrac{1}{6} = \tfrac{1}{4},$$

and hence, by Theorem 2.2.4, the Newton step takes us from x to a point x_1 for which $\|n_{\eta_1}(x_1)\|_{x_1} \leq \tfrac{1}{9}$, putting us precisely in the setting of the earlier analysis (where $\alpha = \tfrac{1}{9}$ was determined safe).

How much will ν have to be decreased from the initial value $\nu = 1$ before we compute a point x for which $\|g_x(x)\|_x^* \leq \tfrac{1}{6}$ so that paths can be switched? An answer is found from the relations

$$\|g_x(x)\|_x = \|\nu g_x(x') + n'_\nu(x)\|_x$$
$$\leq \nu \|g_x(x')\|_x + \|n'_\nu(x)\|_x$$
$$\leq \nu \vartheta_f \left(1 + \frac{1}{\mathrm{sym}(x', D_f)}\right) + \|n'_\nu(x)\|_x,$$

the last inequality by Proposition 2.3.7. In particular, with $\|n'_\nu(x)\|_x \leq \frac{1}{9}$, ν need only satisfy

$$\nu \leq \frac{1}{18 \vartheta_f (1 + 1/\mathrm{sym}(x', D_f))} \tag{2.19}$$

in order for $\|g_x(x)\|_x \leq \frac{1}{6}$.

The requirement on ν specified by (2.19) gives geometric interpretation to the efficiency of the algorithm in following the path $\{z'(\nu) : \nu > 0\}$, beginning with the initial value $\nu = 1$. If the domain D_f is nearly symmetric about the initial point x', not much time will be required to follow the path to a point where we can switch to following the path $\{z(\eta) : \eta > 0\}$.

We stipulated that the algorithm switch paths when it encounters x satisfying $\|g_x(x)\|_x \leq \frac{1}{6}$, and we stipulated that one choose the initial value $\eta_1 := 1/12\|c_x\|_x$. Letting

$$V := \sup\{\langle c, x \rangle : x \in D_f\},$$

note that (2.13) implies

$$\|c_x\|_x \leq V - \mathrm{val},$$

and hence

$$\eta_1 \geq 1/12 \, (V - \mathrm{val}).$$

We have now essentially proven the following theorem.

Theorem 2.4.1. *Assume $f \in \mathcal{SCB}$ and D_f is bounded. Assume $x' \in D_f$, a point at which to initiate the barrier method. If $0 < \epsilon < 1$, then within*

$$O\left(\sqrt{\vartheta_f} \log\left(\frac{\vartheta_f}{\epsilon \, \mathrm{sym}(x', D_f)}\right)\right)$$

iterations of the algorithm, all points x computed thereafter satisfy

$$\frac{\langle c, x \rangle - \mathrm{val}}{V - \mathrm{val}} \leq \epsilon.$$

Consider the following modification to the algorithm. Choose $V' > \langle c, x' \rangle$. Rather than relying on f, rely on the barrier functional

$$x \mapsto f(x) - \ln(V' - \langle c, x \rangle),$$

a functional whose domain is

$$D_f \cap \{x : \langle c, x \rangle < V'\} \tag{2.20}$$

2.4. Primal Algorithms

and whose complexity value does not exceed $\vartheta_f + 1$. In the theorem, the quantity $V - \text{val}$ is then replaced by the potentially much smaller quantity $V' - \text{val}$. Of course the quantity $\text{sym}(x', D_f)$ must then be replaced by the symmetry of the set (2.20) about x'.

Finally, we highlight an implicit assumption underlying our analysis, namely, the complexity value ϑ_f is known. The value is used to safely increase the parameter η. What is actually required is an upper bound $\vartheta \geq \vartheta_f$. If one relies on an upper bound ϑ rather than the precise complexity value ϑ_f, then ϑ_f in the theorem must be replaced by ϑ.

Except for ϑ_f, none of the quantities appearing in the theorem are assumed to be known or approximated. The quantities appear naturally in the analysis of the algorithm, but the algorithm itself does not rely on the quantities.

No ipm's have proven complexity bounds which are better than (2.18) even in the restricted setting of LP. In the case of LP where $\vartheta_f = n$, a bound like (2.18) was first established in [18] for an algorithm other than the barrier method; it was established by Gonzaga [6] for the barrier method. By contrast, in the complexity analysis that started the waves of ipm papers, Karmarkar [10] proved for LP a bound like (2.18) in which the factor $\sqrt{n} \, (= \sqrt{\vartheta_f})$ is replaced by n.

Although no ipm's have proven complexity bounds which are better than (2.18), the barrier method is not considered to be practically efficient relative to some other ipm's, especially relative to primal-dual methods (discussed in sections 3.7 and 3.8). The barrier method is an excellent algorithm with which to begin one's understanding of ipm's, and it is often the perfect choice for concise complexity theory proofs, but it is not one of the ipm's that appear in widely used software.

2.4.3 The Long-Step Barrier Method

One of the barrier method's shortcomings is obvious, being implicit in the terminology "short-step algorithm." Although it is *always* safe to increase η by a factor $1 + 1/8\sqrt{\vartheta_f}$ with each iteration, that increase is small if ϑ_f is large. No doubt, for many instances, a much larger increase is safe.

There is a trivial manner in which to modify the barrier method in hopes of having a more practical algorithm. Rather than increase η by the safe amount, increase it by much more, apply (perhaps several iterations of) Newton's method, and check (say, using Theorem 2.2.5) if the computed point is near the desired minimizer. If not, increase η by a smaller amount and try again.

A more interesting and more practical modification of the barrier method is known as the "long-step barrier method." In this version, one increases η by an arbitrarily large amount but does not take Newton steps. Instead, the Newton steps are used as directions for "exact line searches," as we now describe.

Assume as before that we have an initial value $\eta_1 > 0$ and a point x_1 approximating $z(\eta_1)$. Choose η_2 larger than η_1, perhaps significantly larger. In search of a point x_2 which approximates $z(\eta_2)$, the algorithm will generate a finite sequence of points

$$y_1 := x_1, y_2, \ldots, y_{K-1}, y_K$$

and then let $x_2 := y_K$. At each point y_k, the algorithm will determine if the point is close to $z(\eta_2)$ by, say, checking whether $\|n_{\eta_2}(y_k)\|_{y_k} \leq \frac{1}{4}$. (We choose the specific value $\frac{1}{4}$ because it is the largest value for which Theorem 2.2.5 applies.) The point y_K will be the first point that is determined to satisfy this inequality.

To compute y_{k+1} from y_k, the algorithm minimizes the univariate functional

$$t \mapsto f(y_k + t\, n_{\eta_2}(y_k)). \tag{2.21}$$

This is the place in the algorithm to which the phrase "exact line search" alludes. "Line" refers to the functional being univariate. "Exact" refers to an assumption that the exact minimizer is computed, certainly an exaggeration, but an assumption useful for keeping the analysis succinct. Letting t_{k+1} denote the exact minimizer, define

$$y_{k+1} := y_k + t_{k+1} n_{\eta_2}(y_k),$$

thus ending our description of the long-step barrier method.

The short-step barrier method is confined to making slow but sure progress. The long-step method is more adventurous, having the potential for much quicker progress.

Clearly, the complexity analysis of the long-step barrier method revolves around determining an upper bound on K in terms of the ratio η_2/η_1. We now undertake the task of determining such a bound.

We begin by determining an upper bound on the difference

$$\rho := f_{\eta_2}(x_1) - f_{\eta_2}(z(\eta_2)).$$

Then we show that $f_{\eta_2}(y_k) - f_{\eta_2}(y_{k+1})$ is bounded below by a positive amount τ independent of k; that is, each exact line search decreases the value of f_{η_2} by at least a certain amount. Consequently $K \leq \rho/\tau$. Proofs like this—showing a certain functional decreases by at least a fixed amount with each iteration—are common in the ipm literature.

In proving an upper bound on the difference ρ, we make use of the fact that for any convex functional f and $x, y \in D_f$, one has

$$f(x) - f(y) \leq \langle g(x), x - y \rangle. \tag{2.22}$$

The upper bound on ρ is obtained by adding upper bounds for

$$\rho_1 := f_{\eta_2}(x_1) - f_{\eta_2}(z(\eta_1)) \quad \text{and} \quad \rho_2 := f_{\eta_2}(z(\eta_1)) - f_{\eta_2}(z(\eta_2)).$$

Assuming x_1 is close to $z(\eta_1)$ in the sense that $\|n_{\eta_1}(x_1)\|_{x_1} \leq \frac{1}{4}$, we see that Theorem 2.2.5 implies $\|x_1 - z(\eta_1)\|_{x_1} < \frac{3}{4}$. Thus, applying (2.22) to the functional f_{η_2},

$$\begin{aligned}
\rho_1 &\leq \langle -n_{\eta_2}(x_1), x_1 - z(\eta_1) \rangle_{x_1} \\
&= \tfrac{\eta_2}{\eta_1} \langle n_{\eta_1}(x_1), z(\eta_1) - x_1 \rangle_{x_1} \\
&\quad + \left(\tfrac{\eta_2}{\eta_1} - 1 \right) \langle g_{x_1}(x_1), z(\eta_1) - x_1 \rangle_{x_1} \\
&\leq \tfrac{\eta_2}{\eta_1} \tfrac{1}{4} \tfrac{3}{4} + \left(\tfrac{\eta_2}{\eta_1} - 1 \right) \tfrac{3}{4} \sqrt{\vartheta_f} \\
&\leq \tfrac{\eta_2}{\eta_1} \sqrt{\vartheta_f}.
\end{aligned}$$

2.4. Primal Algorithms

Similarly, for all $y \in D_f$,

$$f_{\eta_2}(z(\eta_1)) - f_{\eta_2}(y) \leq \langle -n_{\eta_2}(z(\eta_1)), z(\eta_1) - y \rangle_{z(\eta_1)}$$
$$= \frac{\eta_2}{\eta_1} \langle n_{\eta_1}(z(\eta_1)), y - z(\eta_1) \rangle_{z(\eta_1)}$$
$$+ \left(\frac{\eta_2}{\eta_1} - 1\right) \langle g_{z(\eta_1)}(z(\eta_1)), y - z(\eta_1) \rangle_{z(\eta_1)}$$
$$= \left(\frac{\eta_2}{\eta_1} - 1\right) \langle g_{z(\eta_1)}(z(\eta_1)), y - z(\eta_1) \rangle_{z(\eta_1)},$$

the final equality because $z(\eta_1)$ minimizes f_{η_1} and hence $n_{\eta_1}(z(\eta_1)) = 0$. Thus,

$$\rho_2 \leq \left(\frac{\eta_2}{\eta_1} - 1\right) \langle g_{z(\eta_1)}(z(\eta_1)), z(\eta_2) - z(\eta_1) \rangle_{z(\eta_1)}$$
$$\leq \frac{\eta_2}{\eta_1} \vartheta_f,$$

the last inequality by Theorem 2.3.3.

Now we show $f_{\eta_2}(y_k) - f_{\eta_2}(y_{k+1})$ is bounded below by a positive amount τ independent of k.

If the algorithm proceeds from y_k to y_{k+1}, it is because y_k happens not to be appropriately close to $z(\eta_2)$, i.e., it happens that $\|n_{\eta_2}(y_k)\|_{y_k} > \frac{1}{4}$. Letting $\tilde{t} := 1/5\|n_{\eta_2}(y_k)\|_{y_k}$ and $\tilde{y} := y_k + \tilde{t} n_{\eta_2}(y_k)$, Theorem 2.2.2 then implies

$$f_{\eta_2}(\tilde{y}) \leq f_{\eta_2}(y_k) - \frac{1}{4}\frac{1}{5} + \frac{1}{2}\left(\frac{1}{5}\right)^2 + \frac{(1/5)^3}{3(1-1/5)}$$
$$\leq f_{\eta_2}(y_k) - \frac{1}{40}.$$

Since t_{k+1} minimizes the functional (2.21), we thus have

$$f_{\eta_2}(y_k) - f_{\eta_2}(y_{k+1}) \geq \tau := \frac{1}{40}.$$

Finally,

$$K \leq \frac{\rho}{\tau} \leq \frac{\rho_1 + \rho_2}{\tau} \leq \frac{40\eta_2}{\eta_1}(\vartheta_f + \sqrt{\vartheta_f}).$$

It follows that if one fixes a constant $\kappa > 1$ and always chooses successive values η_i, η_{i+1} to satisfy $\eta_{i+1} = \kappa \eta_i$, the number of points generated by the long-step barrier method (i.e., the number of exact line searches) in increasing the parameter from an initial value η_1 to some value $\eta > \eta_1$ is

$$O(\kappa \vartheta_f \log_\kappa(\eta/\eta_1)).$$

In the case of linear programming where $\vartheta_f = n$, such a bound was first established by Gonzaga [7] (see also den Hertog, Roos, and Vial [3]).

Fixing κ (say, $\kappa = 100$), we obtain the bound

$$O(\vartheta_f \log(\eta/\eta_1)).$$

This bound is worse than the analogous bound (2.17) for the short-step method by a factor $\sqrt{\vartheta_f}$. It is one of the ironies of the ipm literature that algorithms which are more efficient in practice often have somewhat worse complexity bounds.

2.4.4 A Predictor-Corrector Method

The Newton step $n_\eta(x) := -\eta c_x - g_x(x)$ for the barrier method can be viewed as the sum of two steps, one of which predicts the tangential direction of the central path and the other of which corrects for the discrepancy between the tangential direction and the actual position of the (curving) path.

The corrector step is the Newton step at x for the barrier functional $f|_{L(v)}$ where $v = \langle c, x \rangle$ and
$$L(v) := \{y : \langle c, y \rangle = v\}.$$
Thus, the corrector step is $n|_{L(v)}(x)$, this being the orthogonal projection of the Newton step $n(x)$ for f onto the subspace $L(0) = \{y : \langle c, y \rangle = 0\}$, orthogonal w.r.t. $\langle\ ,\ \rangle_x$. In the literature, the corrector step is often referred to as the "centering direction." It aims to move from x toward the point on the central path having the same objective value as x.

Since the multiples of c_x $(= H(x)^{-1}c)$ form the orthogonal complement of $L(0)$, the difference $n(x) - n|_{L(v)}(x)$ is a multiple of c_x, and hence so is the predictor step
$$n_\eta(x) - n|_{L(v)}(x) = -\eta c_x + (n(x) - n|_{L(v)}(x)).$$

The vector $-c_x$ predicts the tangential direction of the central path near x. If x is on the central path, the vector $-c_x$ is exactly tangential to the path, pointing in the direction of decreasing objective values. (Indeed, observe that by differentiating both sides of the identity $\eta c + g(z(\eta)) = 0$ w.r.t. η, we have $c + H(z(\eta))z'(\eta) = 0$, i.e., $z'(\eta) := -c_{z(\eta)}$.) In the literature, $-c_x$ is often referred to as the "affine-scaling direction." With regards to $\langle\ ,\ \rangle_x$, it is the direction in which one would move to decrease the objective value most quickly.

Whereas the barrier method combines a predictor step and a corrector step in one step, predictor-corrector methods separate the two types of steps. After a predictor step, several corrector steps might be applied. In practice, predictor-corrector methods tend to be substantially more efficient than the barrier method, but the (worst-case) complexity bounds that have been proven for them are worse.

Perhaps the most natural predictor-corrector method is based on moving in the predictor direction a fixed fraction of the distance toward the boundary and then recentering via exact line searches. We now formalize and analyze such an algorithm.

Fix σ satisfying $0 < \sigma < 1$. Assume x_1 is near the central path. Let $v_1 := \langle c, x_1 \rangle$. The algorithm first computes
$$s_1 := \sup\{s : x_1 - sc_{x_1} \in D_f\}.$$
Let $y_1 := x_1 - \sigma s_1 c_{x_1}$. Thus, $-\sigma s_1 c_{x_1}$ is the predictor step. Let
$$v_2 := \langle c, y_1 \rangle = v_1 - \sigma s_1 \|c_{x_1}\|_{x_1}^2.$$

Beginning with y_1, the algorithm takes corrector steps, moving toward the point z_2 on the central path with objective value v_2 by using the Newton steps for the functional $f|_{L(v_2)}$ as directions in performing exact line searches. Precisely, given y_k, the algorithm computes the minimizer t_{k+1} for the univariate functional
$$t \mapsto f(y_k + tn|_{L(v_2)}(y_k)).$$

2.4. Primal Algorithms

Let
$$y_{k+1} := y_k + t_{k+1} n|_{L(v_2)}(y_k).$$

When the first point y_K is encountered for which $\|n|_{L(v_2)}(y_K)\|_{y_K}$ is appropriately small, the algorithm lets $x_2 := y_K$ and takes a predictor step from x_2, relying on the same value σ as in the predictor step from x_1. The predictor step is followed by corrector steps, and so on.

Typically, σ is chosen near 1, say, $\sigma = .99$. In the following analysis, we assume $\sigma \geq \frac{1}{4}$.

In analyzing the predictor-corrector method, we determine an upper bound on the number K of exact line searches made in moving from x_1 to x_2, and we determine a lower bound on progress made in decreasing the objective value by moving from x_1 to x_2.

Unlike the previous algorithms, the predictor-corrector method does not rely on the parameter η. However, it is useful to rely on η in analyzing the method so as to make use of the previous analysis. In this regard, let η_1 be the (positive) value satisfying $\langle c, z(\eta_1) \rangle = v_1$, i.e., $z(\eta_1)$ is the point on the central path whose objective value is the same as the objective value for x_1.

For the analysis of the predictor-corrector method, we consider $\|n|_{L(v)}(x)\|_x \leq \frac{1}{14}$ to be the criterion for claiming x to be close to the point z on the central path satisfying $\langle c, z \rangle = v (= \langle c, x \rangle)$. The specific value $\frac{1}{14}$ is chosen so that we are in position to rely on the barrier method analysis. Specifically, we claim that $\|n|_{L(v_1)}(x_1)\|_{x_1} \leq \frac{1}{14}$ implies $\|n_{\eta_1}(x_1)\|_{x_1} \leq \frac{1}{9}$, precisely the criterion we assumed x_1 to satisfy for the barrier method. To justify the claim, let z_1 denote the minimizer of $f|_{L(v_1)}$; thus, $z_1 = z(\eta_1)$. If $\|n|_{L(v_1)}(x_1)\|_{x_1} \leq \frac{1}{14}$, then Theorem 2.2.5 applied to $f|_{L(v_1)}$ implies

$$\|z_1 - x_1\|_{x_1} \leq \frac{1}{11}. \tag{2.23}$$

Consequently, since $z_1 = z(\eta_1)$, applying Theorem 2.2.3 to f_{η_1} yields

$$\|n_{\eta_1}(x_1)\|_{x_1} \leq \|z_1 - x_1\|_{x_1} + \frac{\|z_1 - x_1\|_{x_1}^2}{1 - \|z_1 - x_1\|_{x_1}} \leq \frac{1}{9}.$$

The barrier method moves from x_1 to $x_1 + n_{\eta_2}(x_1)$, where $\eta_2 = (1 + 1/8\sqrt{\vartheta_f})\eta_1$. The length of the barrier method step is thus

$$\|n_{\eta_2}(x_1)\|_{x_1} = \left\| \frac{\eta_2}{\eta_1} n_{\eta_1}(x_1) + \left(\frac{\eta_2}{\eta_1} - 1\right) g_{x_1}(x_1) \right\|_{x_1}$$
$$\leq \frac{9}{8} \frac{1}{9} + \frac{1}{8}$$
$$= \frac{1}{4}.$$

In taking the step, the barrier method decreases the objective function value from $\langle c, x_1 \rangle$ to $\langle c, x_1 + n_{\eta_2}(x_1) \rangle$; hence, the decrease is at most

$$|\langle c, n_{\eta_2}(x_1) \rangle| = |\langle c_{x_1}, n_{\eta_2}(x_1) \rangle_{x_1}| \leq \frac{1}{4} \|c_x\|_{x_1}.$$

Assuming $\sigma \geq \frac{1}{4}$ for the predictor-corrector method, the predictor step is in the direction $-c_{x_1}$ and has length at least $\frac{1}{4}$, a consequence of $B_{x_1}(x_1, 1) \subseteq D_f$. Thus, $v_1 - v_2 \geq \frac{1}{4} \|c_{x_1}\|_{x_1}$. Hence, in moving from x_1 to x_2, the progress made by the predictor-corrector method in decreasing the objective value is at least as great as the progress made by the barrier method in taking one step from x_1. (Clearly, the point x_2 computed by the predictor-corrector method generally differs from the point x_2 computed by the barrier method.) Of course in moving from x_1 to x_2, the predictor-corrector method might require several exact line searches. We now bound the number K of exact line searches.

Analogous to our analysis for the long-step barrier method, we obtain an upper bound on K by dividing an upper bound on $f(y_1) - f(z_2)$ by a lower bound on the differences $f(y_k) - f(y_{k+1})$.

The lower bound on the differences $f(y_k) - f(y_{k+1})$ is proven exactly, as was the lower bound for the differences $f_{\eta_2}(y_k) - f_{\eta_2}(y_{k+1})$ in our analysis of the long-step barrier method. Assuming $\|n|_{L(v_2)}(y_k)\|_{y_k} > \frac{1}{14}$ (as is the case if the algorithm proceeds to compute y_{k+1}), one relies on Theorem 2.2.2, now applied to the functional $f|_{L(v_2)}$, to show $f(y_k) - f(y_{k+1})$ is bounded below independently of k.

To obtain an upper bound on $f(y_1) - f(z_2)$, one can use the identity

$$f(y_1) - f(z_2) = (f(y_1) - f(x_1)) + (f(x_1) - f(z_1)) + (f(z_1) - f(z_2)).$$

Theorem 2.3.8 and the definition of y_1 imply

$$f(y_1) - f(x_1) \le -\vartheta_f \ln(1 - \sigma).$$

Relation (2.22) applied to $f|_{L(v_1)}$, together with (2.23), gives

$$f(x_1) - f(z_1) \le \|n|_{L(v_1)}(x_1)\|_{x_1} \|z_1 - x_1\|_{x_1} \le \frac{1}{14}\frac{1}{11} = \frac{1}{154}.$$

Finally, (2.22) applied to f gives

$$f(z_1) - f(z_2) \le \langle -\eta_1 c, z_1 - z_2 \rangle = -\eta_1(v_1 - v_2) \le 0.$$

In all,

$$f(y_1) - f(z_2) \le \vartheta_f \ln(\tfrac{1}{1-\sigma}) + \tfrac{1}{154}.$$

Combined with the constant lower bound on the differences $f(y_k) - f(y_{k+1})$, we thus find that the number K of exact line searches performed in moving from x_1 to x_2 satisfies

$$K = O\left(\vartheta_f \log\left(\frac{1}{1-\sigma}\right)\right).$$

Having shown the progress made by the predictor-corrector method in decreasing the objective value is at least as great as the progress made by the barrier method in taking one step from x_1, we obtain complexity bounds for the predictor-corrector method which are greater than the bounds for the barrier method by a factor K, that is, by a factor ϑ_f (assuming σ fixed, say, $\sigma = .99$). The bounds are greater than the bounds for the long-step barrier method by a factor $\sqrt{\vartheta_f}$.

2.5 Matters of Definition

There are various equivalent ways of defining self-concordant functionals. Our definition is geometric and simple to employ in theory, but it is not the original definition due to

2.5. Matters of Definition

Nesterov and Nemirovskii [15]. In this section we consider various equivalent definitions of self-concordance, including the original definition. We close the section with a brief discussion of the term "strongly nondegenerate," a term we have suppressed.

The proofs in this section are more technical than conceptual, yielding useful results, but ones that are not central to understanding the core theory of ipm's. It is suggested that on a first pass through the book, the reader absorb only the exposition and statements of results, bypassing the formal proofs.

Unless otherwise stated, we assume only that $f \in C^2$, D_f is open and convex, and $H(x)$ is pd for all $x \in D_f$.

For ease of reference, we recall our definition of self-concordance.

A functional f is said to be *(strongly nondegenerate) self-concordant* if for all $x \in D_f$, we have $B_x(x, 1) \subseteq D_f$, and if whenever $y \in B_x(x, 1)$, we have

$$1 - \|y - x\|_x \leq \frac{\|v\|_y}{\|v\|_x} \leq \frac{1}{1 - \|y - x\|_x} \quad \text{for all } v \neq 0.$$

Recall $\mathcal{SC}$ denotes the family of functionals thus defined.

An important property in establishing the equivalence of various definitions of self-concordance is the "transitivity" of the condition

$$\text{for all } v \neq 0, \quad \frac{\|v\|_y}{\|v\|_x} \leq \frac{1}{1 - \|y - x\|_x}. \tag{2.24}$$

Specifically, if x, y, and z are colinear with y between x and z, if x and y satisfy (2.24), if y and z satisfy the analogous condition

$$\text{for all } v \neq 0, \quad \frac{\|v\|_z}{\|v\|_y} \leq \frac{1}{1 - \|z - y\|_y}, \tag{2.25}$$

and if $\|z - x\|_x < 1$, then

$$\text{for all } v \neq 0, \quad \frac{\|v\|_z}{\|v\|_x} \leq \frac{1}{1 - \|z - x\|_x}. \tag{2.26}$$

To establish this transitivity, note (2.24) implies

$$\|z - y\|_y \leq \frac{\|z - y\|_x}{1 - \|y - x\|_x}.$$

Substituting into (2.25) gives for all $v \neq 0$,

$$\frac{\|v\|_z}{\|v\|_y} \leq \frac{1 - \|y - x\|_x}{1 - \|y - x\|_x - \|z - y\|_x}$$

$$= \frac{1 - \|y - x\|_x}{1 - \|z - x\|_x}, \tag{2.27}$$

the equality relying on the colinearity of x, y, and z. Note (2.26) is immediate from (2.24) and (2.27), hence the transitivity.

We now show that in the definition of self-concordance, the lower bound on $\|v\|_y/\|v\|_x$ is redundant.

Theorem 2.5.1. *Assume f is such that for all $x \in D_f$ we have $B_x(x, 1) \subseteq D_f$ and is such that whenever $y \in B_x(x, 1)$ we have*

$$\frac{\|v\|_y}{\|v\|_x} \leq \frac{1}{1 - \|y - x\|_x} \text{ for all } v \neq 0. \tag{2.28}$$

Then

$$1 - \|y - x\|_x \leq \frac{\|v\|_y}{\|v\|_x} \text{ for all } v \neq 0.$$

Proof. Since

$$\|H_x(y)^{-1}\|_x = \frac{1}{\inf_{v \neq 0} \|v\|_y^2/\|v\|_x^2} = \sup_{v \neq 0} \frac{\|v\|_x^2}{\|v\|_y^2},$$

it suffices to show for all positive ϵ in some open neighborhood of 0, and for all $x, y \in D_f$, that

$$\|y - x\|_x < \frac{1}{1+\epsilon} \Rightarrow \|H_x(y)^{-1}\|_x \leq \frac{1}{(1 - (1+\epsilon)\|y - x\|_x)^2}. \tag{2.29}$$

Toward establishing the implication (2.29), note from (2.28) that whenever points x and y satisfy $\|y - x\|_x \leq \frac{\epsilon}{1+\epsilon}$ (with ϵ small), we have

$$\|y - x\|_y \leq \frac{\|y - x\|_x}{1 - \|y - x\|_x} \leq (1+\epsilon)\|y - x\|_x. \tag{2.30}$$

Letting $\lambda_{\min}$ denote the minimum eigenvalue of $H_x(y)$, from the identities $H_x(y) = H(x)^{-1}H(y) = H_y(x)^{-1}$ we of course have

$$\|H_x(y)^{-1}\|_x = 1/\lambda_{\min} = \|H_y(x)\|_y.$$

Since

$$\|H_y(x)\|_y = \sup_{v \neq 0} \frac{\|v\|_x^2}{\|v\|_y^2},$$

applying (2.28) with x and y interchanged thus gives

$$\|H_x(y)^{-1}\|_x = \|H_y(x)\|_y \leq \frac{1}{(1 - \|x - y\|_y)^2}.$$

Hence, by (2.30),

$$\|H_x(y)^{-1}\|_x \leq \frac{1}{(1 - (1+\epsilon)\|x - y\|_x)^2}.$$

Summarizing, for ϵ in an open neighborhood of 0, and for all $x, y \in D_f$,

$$\|y - x\|_x \leq \frac{\epsilon}{1+\epsilon} \Rightarrow \|H_x(y)^{-1}\|_x \leq \frac{1}{(1 - (1+\epsilon)\|y - x\|_x)^2}.$$

This gives the desired implication (2.29) except in that $\|y - x\|_x \leq \frac{\epsilon}{1+\epsilon}$ replaces $\|y - x\|_x < \frac{1}{1+\epsilon}$. The proof strategy now is to show that when one has the desired implication for all

2.5. Matters of Definition

$x, y \in D_f$ except in that y is required to lie in a ball with smaller radius r than the claimed radius (i.e., $\|y - x\|_x < r$ rather than $\|y - x\|_x < \frac{1}{1+\epsilon}$, where r is independent of x, y), then the smaller radius r can be increased to $r' := r + \epsilon(\frac{1}{1+\epsilon} - r)$. Likewise, r' can be increased. In the limit, we arrive at the validity of desired implication (2.29) for the claimed radius $\frac{1}{1+\epsilon}$.

To implement this proof strategy, assume $0 < \epsilon < 1$, $\frac{\epsilon}{1+\epsilon} \leq r < \frac{1}{1+\epsilon}$, and assume that for all $x, y \in D_f$,

$$\|y - x\|_x < r \implies \|H_x(y)^{-1}\|_x \leq \frac{1}{(1 - (1+\epsilon)\|y - x\|_x)^2}. \tag{2.31}$$

Define $z := y + t(y - x)$, where t is chosen so that

$$\|z - y\|_x = \epsilon \left(\frac{1}{1+\epsilon} - \|y - x\|_x\right).$$

Since x and y are arbitrary, to prove that r can be increased to r' it suffices to show that (2.31) gives

$$\|H_x(z)^{-1}\|_x \leq \frac{1}{(1 - (1+\epsilon)\|z - x\|_x)^2}. \tag{2.32}$$

Using (2.28) with $v = z - y$ and the definition of z, we see

$$\|z - y\|_y \leq \frac{\|z - y\|_x}{1 - \|y - x\|_x} \tag{2.33}$$

$$= \frac{\frac{\epsilon}{1+\epsilon}(1 - (1+\epsilon)\|y - x\|_x)}{1 - \|y - x\|_x}$$

$$\leq \frac{\epsilon}{1+\epsilon}.$$

Hence, by (2.31) applied to y and z (rather than to x and y),

$$\|H_y(z)^{-1}\|_y \leq \frac{1}{(1 - (1+\epsilon)\|z - y\|_y)^2}. \tag{2.34}$$

Trivially, by (2.33) we have

$$\|z - y\|_y \leq \frac{\|z - y\|_x}{1 - \|y - x\|_x} \leq \frac{\|z - y\|_x}{1 - (1+\epsilon)\|y - x\|_x}.$$

Substituting this in (2.34) gives

$$\|H_y(z)^{-1}\|_y \leq \left(\frac{1 - (1+\epsilon)\|y - x\|_x}{1 - (1+\epsilon)(\|z - y\|_x + \|y - x\|_x)}\right)^2$$

$$= \left(\frac{1 - (1+\epsilon)\|y - x\|_x}{1 - (1+\epsilon)\|z - x\|_x}\right)^2,$$

the equality relying on the colinearity of x, y, and z. That is, for all $v \neq 0$,

$$\frac{\|v\|_y}{\|v\|_z} \leq \frac{1 - (1+\epsilon)\|y - x\|_x}{1 - (1+\epsilon)\|z - x\|_x}.$$

If we assume
$$\frac{\|v\|_x}{\|v\|_y} \leq \frac{1}{1-(1+\epsilon)\|y-x\|_x}, \qquad (2.35)$$
we thus have
$$\frac{\|v\|_x}{\|v\|_z} \leq \frac{1}{1-(1+\epsilon)\|z-x\|_x}$$
and hence the relation (2.32). Since (2.35) follows from the bound
$$\|H_x(y)^{-1}\|_x \leq \frac{1}{(1-(1+\epsilon)\|y-x\|_x)^2},$$
we have thus shown the implication (2.31) to indeed give the relation (2.32), thereby completing the proof. □

The following theorem provides various equivalent definitions of self-concordant functionals.

Theorem 2.5.2. *The following conditions on a functional f are equivalent:*

(1a) *For all $x, y \in D_f$, if $\|y-x\|_x < 1$, then*
$$\frac{\|v\|_y}{\|v\|_x} \leq \frac{1}{1-\|y-x\|_x} \quad \text{for all } v \neq 0.$$

(1b) *For each $x \in D_f$, and for all y in some open neighborhood of x,*
$$\frac{\|v\|_y}{\|v\|_x} \leq \frac{1}{1-\|y-x\|_x} \quad \text{for all } v \neq 0.$$

(1c) *For all $x \in D_f$,*
$$\limsup_{y \to x} \frac{\|I - H_x(y)\|_x}{\|y-x\|_x} \leq 2.$$

Moreover, if f satisfies any (and hence all) of the above conditions, as well as any condition from the following list, then f satisfies all conditions from the following list:

(2a) *For all $x \in D_f$ we have $B_x(x, 1) \subseteq D_f$.*

(2b) *There exists $0 < r \leq 1$ such that for all $x \in D_f$ we have $B_x(x, r) \subseteq D_f$.*

(2c) *If a sequence $\{x_k\}$ converges to a point in the boundary ∂D_f, then $f(x_k) \to \infty$.*

Hence, since SC consists precisely of those functionals satisfying conditions 1a and 2a, by choosing one condition from the first set and one from the second, we can define the set SC as the set of functionals satisfying the two chosen conditions.

Proof. To prove the theorem, we first establish the equivalence of conditions 1a, 1b, and 1c. We then prove conditions 1a and 2b together imply 2a, as do conditions 1a and 2c. Trivially, 2a implies 2b. To conclude the proof, it then suffices to recall that by Theorem 2.2.9, conditions 1a and 2a together imply 2c.

2.5. Matters of Definition

Now we establish the equivalence of 1a, 1b, and 1c. Trivially, 1a implies 1b. Next note

$$\|I - H_x(y)\|_x = \max_{v \neq 0} \left| \frac{\langle v, [I - H_x(y)]v \rangle_x}{\|v\|_x^2} \right|$$

$$= \max_{v \neq 0} \left| 1 - \frac{\|v\|_y^2}{\|v\|_x^2} \right|.$$

Consequently, condition 1b implies for y near x,

$$\|I - H_x(y)\|_x \leq \frac{1}{(1 - \|y - x\|_x)^2} - 1$$

$$= \frac{2\|y - x\|_x - \|y - x\|_x^2}{(1 - \|y - x\|_x)^2}.$$

It easily follows that 1b implies 1c.

To conclude the proof of the equivalence of 1a, 1b, and 1c, we assume 1c holds but 1a does not, then obtain a contradiction.

Since

$$\|H_x(y)\|_x = \max_{v \neq 0} \frac{\|v\|_y^2}{\|v\|_x^2},$$

condition 1a not holding implies there exist x, y', and $\epsilon > 0$ such that $\|y' - x\|_x < \frac{1}{1+\epsilon}$ and

$$\|H_x(y')\|_x > \frac{1}{(1 - (1+\epsilon)\|y' - x\|_x)^2}.$$

Considering points on the line segment between x and y', and relying on continuity of the Hessian, it then readily follows that there exists y (possibly $y = x$) satisfying $\|y - x\|_x < \frac{1}{1+\epsilon}$,

$$\|H_x(y)\|_x = \frac{1}{(1 - (1+\epsilon)\|y - x\|_x)^2}, \tag{2.36}$$

and

$$\|H_x(z)\|_x > \frac{1}{(1 - (1+\epsilon)\|z - x\|_x)^2} \tag{2.37}$$

for all $z := y + t(y - x)$, where $t > 0$ is sufficiently small.

Condition 1c implies for z near y,

$$\|H_y(z)\|_y \leq \frac{1}{(1 - (1+\epsilon)\|z - y\|_y)^2}.$$

That is, for all $v \neq 0$,

$$\frac{\|v\|_z}{\|v\|_y} \leq \frac{1}{1 - (1+\epsilon)\|z - y\|_y}. \tag{2.38}$$

Likewise, from (2.36),

$$\frac{\|v\|_y}{\|v\|_x} \leq \frac{1}{1 - (1+\epsilon)\|y - x\|_x}. \tag{2.39}$$

In particular,
$$\|z-y\|_y \le \frac{\|z-y\|_x}{1-(1+\epsilon)\|y-x\|_x}.$$

Substituting into (2.38) and relying on the colinearity of x, y, and z, we have
$$\frac{\|v\|_z}{\|v\|_y} \le \frac{1-(1+\epsilon)\|y-x\|_x}{1-(1+\epsilon)\|z-x\|_x}. \tag{2.40}$$

From (2.39) and (2.40), for all $v \ne 0$,
$$\frac{\|v\|_z}{\|v\|_x} \le \frac{1}{1-(1+\epsilon)\|z-x\|_x},$$

that is,
$$\|H_x(z)\|_x \le \frac{1}{(1-(1+\epsilon)\|z-x\|_x)^2},$$

contradicting (2.37). We have thus proven the equivalence of conditions 1a, 1b, and 1c.

Now we prove that conditions 1a and 2b together imply 2a. Let $0 < r \le 1$ be as in 2b, i.e., $B_x(x,r) \subseteq D_f$ for all $x \in D_f$. Let $r' := (2-r)r$. We show 1a and 2b together imply $B_x(x,r') \subseteq D_f$ for all $x \in D_f$, i.e., 2b holds with r' in place of the smaller value r. Likewise, r' can be replaced with a larger value. In the limit we find $B_x(x,1) \subseteq D_f$ for all $x \in D_f$, precisely condition 2a.

To verify that 1a and 2b together allow r to be replaced with the larger value r', assume $x, y \in D_f$ satisfy $\|y-x\|_x < r$. Let
$$z := y + (1 - \|y-x\|_x)(y-x).$$

The colinear points x, y, z satisfy $\|z-x\|_x = (2 - \|y-x\|_x)\|y-x\|_x$. Consequently, since y is an arbitrary point in $B_x(x,r)$, it suffices to verify that 1a and 2b together imply $z \in D_f$.

By condition 2b, $y \in D_f$. Condition 1a applied with $v := z - y$ thus gives
$$\|z-y\|_y \le \frac{\|z-y\|_x}{1-\|y-x\|_x}.$$

Hence, since $\|z-y\|_x = (1-\|y-x\|_x)\|y-x\|_x$ we have $\|z-y\|_y \le \|y-x\|_x < r$. Consequently, 2b can be applied to y and z (in place of x and y), yielding the desired inclusion $z \in D_f$.

To conclude the proof of the theorem, it remains only to prove that conditions 1a and 2c together imply 2a.

Assuming $y \in D_f$ satisfies $\|y-x\|_x < 1$, condition 1a implies
$$f(y) = f(x) + \langle g_x(x), y-x \rangle_x + \int_0^1 \int_0^t \langle y-x, H_x(x+s(y-x))(y-x) \rangle_x \, ds \, dt$$
$$\le f(x) + \|g_x(x)\|_x + \int_0^1 \int_0^t \|H_x(x+s(y-x))\|_x \, ds \, dt$$
$$\le f(x) + \|g_x(x)\|_x + \frac{1}{2(1-\|y-x\|_x)^2}.$$

2.5. Matters of Definition

In particular, $f(y)$ is bounded away from ∞ on each set $B_x(x, r) \cap D_f$, where $r < 1$. By condition 2c we conclude $B_x(x, r) \subseteq D_f$ if $r < 1$. Hence, $B_x(x, 1) \subseteq D_f$, completing the proof. $\square$

We turn to the original definition of self-concordance due to Nesterov and Nemirovskii. First, we provide some motivation.

We know that if one restricts a self-concordant functional f to subspaces—or translates thereof—one obtains self-concordant functionals. In particular, if f is restricted to a line $t \mapsto x + td$ (where $x, d \in \mathbb{R}^n$), then

$$\phi(t) := f(x + td)$$

is a univariate self-concordant functional. Since for ϕ we have

$$\|v\|_t = \sqrt{\phi''(t)}|v|,$$

the property

$$\frac{\|v\|_s}{\|v\|_t} \leq \frac{1}{1 - \|s - t\|_t}$$

is identical to

$$\frac{\sqrt{\phi''(s)}}{\sqrt{\phi''(t)}} \leq \frac{1}{1 - \sqrt{\phi''(t)}|s - t|}.$$

Squaring both sides, then subtracting 1 from both sides, and finally multiplying both sides by $\phi''(t)/|s - t|$, we find

$$\frac{\phi''(s) - \phi''(t)}{|s - t|} \leq \frac{2\phi''(t)^{3/2} - \phi''(t)^2|s - t|}{(1 - \sqrt{\phi''(t)}|s - t|)^2}.$$

If f—and hence ϕ—is thrice-differentiable, we thus have

$$\phi'''(t) \leq 2\phi''(t)^{3/2}. \tag{2.41}$$

This result has a converse. The converse, given by the following theorem, coincides with the original definition of self-concordance due to Nesterov and Nemirovskii. (Keep in mind our standing assumptions that $f \in C^2$, D_f is open and convex, and $H(x)$ is pd for all $x \in D_f$.)

Theorem 2.5.3. *Assume $f \in C^3$ and assume each of the univariate functionals ϕ obtained by restricting f to lines intersecting D_f satisfy (2.41) for all t in their domains. Furthermore, assume that if a sequence $\{x_k\}$ converges to a point in the boundary ∂D_f, then $f(x_k) \to \infty$. Then $f \in \mathcal{SC}$.*

Proof. The proof assumes the reader to be familar with certain properties of differentials.

To prove the theorem, it suffices to prove f satisfies conditions 1c and 2c of Theorem 2.5.2. Of course 2c is satisfied by assumption.

The family of inequalities (2.41) (an inequality for each x, d, and t) is equivalent to the family of inequalities

$$|D^3(x)[u, u, u]| \leq 2 \quad \text{whenever } \|u\|_x \leq 1 \tag{2.42}$$

(an inequality for each x and u). On the other hand, the family of inequalities given by condition 1c is equivalent to

$$|D^3(x)[u, v, v]| \leq 2 \text{ whenever } \|u\|_x, \|v\|_x \leq 1. \tag{2.43}$$

However, for any C^3-functional f and for any inner product norm $\| \ \| := \langle \ , \ \rangle^{1/2}$,

$$\max\{|D^3(x)[u, v, w]| : \|u\|, \|v\|, \|w\| \leq 1\} = \max\{|D^3(x)[u, u, u]| : \|u\| \leq 1\}.$$

Hence, (2.43) follows from (2.42). □

The original definition of (strongly nondegenerate) self-concordance was that f satisfy the assumptions of Theorem 2.5.3. The theorem shows such f to be self-concordant according to our definition. Our definition is ever-so-slightly less restrictive by requiring only $f \in C^2$, not $f \in C^3$. For example, letting $\| \ \|$ denote the Euclidean norm, the functionals

$$f(x) := \tfrac{1}{2}\|x\|^2 + \tfrac{1}{6}\|x\|^3$$

and

$$f(x) := e^{\|x\|} - \|x\|$$

are self-concordant according to our definition but not according to the original definition. Neither functional is thrice-differentiable at the origin.

Our definition was not chosen for the slightly broader set of functionals it defines. It was chosen because it provides the reader with a sense of the geometry underlying self-concordance and because it is handy in developing the theory. Nonetheless, the original definition has distinct advantages, especially in proving a functional to be self-concordant. For example, assume $D \subseteq \mathbb{R}^n$ is open and convex, and assume $F \in C^3$ is a functional which takes on only positive values in D and only the value 0 on the boundary ∂D. Furthermore, assume that for each line intersecting D, the univariate functional $t \mapsto F(x + td)$ obtained by restricting F to the line happens to be a polynomial—moreover, a polynomial with only real roots. Then, relying on the original definition of self-concordance, it is easy to prove that the functional

$$f(x) := -\ln(F(x))$$

is self-concordant. Indeed, letting $\phi(t) := f(x + td)$ be the restriction of f to a line, and letting $r_1, \ldots, r_d$ denote the roots of the univariate polynomial $t \mapsto F(x + td)$ (listed according to multiplicity), we have

$$\phi'''(t) = -\tfrac{d^3}{dt^3} \ln \prod_i (t - r_i)$$

$$= -2 \sum_i \frac{1}{(t - r_i)^3}$$

$$\leq 2 \left(\sum_i \frac{1}{(t - r_i)^2} \right)^{3/2}$$

$$= 2\phi''(t)^{3/2},$$

the inequality due to the relation $\| \ \|_3 \leq \| \ \|_2$ between the 2-norm and the 3-norm on $\mathbb{R}^d$.

2.5. Matters of Definition

It is an insightful exercise to show that the self-concordance of the various logarithmic barrier functions are special cases of the result described in the preceding paragraph. Incidentally, functionals F as above are known as "hyperbolic polynomials" (cf. [2], [9]).

We close this section with a discussion of the qualifying phrase "strongly nondegenerate," which we suppress throughout the book.

Nesterov and Nemirovskii define *self-concordant functionals* as functionals $f \in C^3$, with open and convex domains, satisfying (2.41) for the univariate functionals ϕ obtained by restricting f to lines. It is readily proven that self-concordant functionals (thus defined) have psd Hessians. They define *strongly* self-concordant functionals as having the additional property that $f(x_k) \to \infty$ if the sequence $\{x_k\}$ converges to a point in the boundary ∂D_f. Finally, strongly *nondegenerate* self-concordant functionals are those which satisfy the yet further property that $H(x)$ is pd for all $x \in D_f$.

One might ask if the Nesterov–Nemirovskii definition of, say, strong self-concordance has a geometric analogue similar to our definition of (strongly nondegenerate) self-concordance. One should not expect the analogue to be as intuitively simple as our definition—for the bilinear forms

$$\langle u, v \rangle_x := \langle u, H(x)v \rangle$$

are not inner products if f is not nondegenerate. However, there is indeed an analogue, obtained as a simple generalization of the definition for strongly nondegenerate self-concordance. Roughly, strongly self-concordant functionals are those obtained by extending strongly nondegenerate self-concordant functionals to larger vector spaces by having the functional be constant on parallel slices. Specifically, one can prove (as is done in [15]) that f is strongly self-concordant iff $\mathbb{R}^n$ is a direct sum $L_1 \oplus L_2$ of subspaces for which there exists a strongly nondegenerate self-concordant functional h, with $D_h \subseteq L_1$, satisfying $f(x_1, x_2) = h(x_1)$. For example, $f(x) := -\ln(x_1)$ is a strongly self-concordant functional with domain the half-space $\mathbb{R}_+ \oplus \mathbb{R}^{n-1}$ in $\mathbb{R}^n$, but it is not nondegenerate.

If self-concordant (resp., strongly self-concordant) functionals are added, the resulting functional is self-concordant (resp., strongly self-concordant). If one of the summands is strongly nondegenerate, so is the sum. This indicates how the theory of self-concordant functionals, and strongly self-concordant functionals, parallels the theory developed in this book. To get to the heart of the ipm theory quickly and cleanly, we focus on strongly nondegenerate self-concordant functionals.

Henceforth, we return to our practice of referring to functionals as self-concordant when, strictly speaking, we mean the functionals to be strongly nondegenerate self-concordant.

Chapter 3
Conic Programming and Duality

3.1 Conic Programming

Linear programming (LP) and semidefinite programming (SDP) are special cases of conic programming (CP). The ingredients for a CP instance are a closed, convex cone $K \subseteq \mathbb{R}^n$ (if $x_1, x_2 \in K$ and $t_1, t_2 \geq 0$, then $t_1 x_1 + t_2 x_2 \in K$), a linear operator $A : \mathbb{R}^n \to \mathbb{R}^m$, vectors $c \in \mathbb{R}^n$ and $b \in \mathbb{R}^m$, and for our development, an inner product $\langle \, , \, \rangle$ on $\mathbb{R}^n$. These ingredients give the following instance:

$$\begin{aligned} \min \quad & \langle c, x \rangle \\ \text{s.t.} \quad & Ax = b, \\ & x \in K. \end{aligned} \quad (3.1)$$

LP is obtained with $K = \mathbb{R}^n_+$ (the nonnegative orthant), whereas SDP is obtained by letting $K = \mathbb{S}^{n \times n}_+$ (the cone of psd matrices). If one lets K be the *second-order cone*, that is,

$$K := \left\{ x \in \mathbb{R}^n : \sum_{i=1}^{n-1} x_i^2 \leq x_n^2 \text{ and } x_n \geq 0 \right\},$$

then one has *second-order programming*.

Strictly speaking, "min" should be replaced with "inf" in (3.1) even though K is closed. (We discuss this later.) Let val denote the optimal objective value,

$$\text{val} := \inf\{\langle c, x \rangle : Ax = b \text{ and } x \in K\}.$$

If the constraints are infeasible (i.e., inconsistent), define val $:= \infty$.

We refer to (3.1) as the *primal instance*. The dual of the primal instance is itself a CP instance, its cone being the *dual cone* of K,

$$K^* := \{s \in \mathbb{R}^n : \langle x, s \rangle \geq 0 \text{ for all } x \in K\}.$$

Assuming $\mathbb{R}^m$, like $\mathbb{R}^n$, is endowed with an inner product, the *dual instance* is

$$\begin{aligned} \max \quad & \langle b, y \rangle \\ \text{s.t.} \quad & c - A^* y \in K^*, \end{aligned}$$

where A^* denotes the adjoint of A. The constraints of the dual instance can be written in the same form as the primal if we introduce slack variables $s \in \mathbb{R}^n$:

$$\begin{aligned} \max \quad & \langle b, y \rangle \\ \text{s.t.} \quad & A^*y + s = c, \\ & s \in K^*. \end{aligned}$$

Let val* denote the optimal value of the dual instance: $-\infty$ if the instance is infeasible.

An especially important relation between feasible points for the primal and dual instances is given by the identities

$$\begin{aligned} \langle x, s \rangle &= \langle x, c - A^*y \rangle \\ &= \langle c, x \rangle - \langle Ax, y \rangle \\ &= \langle c, x \rangle - \langle b, y \rangle. \end{aligned} \quad (3.2)$$

Since $x \in K$ and $s \in K^*$ it follows that val* $\leq$ val, an inequality known as *weak duality*.

To understand how the geometry of the primal and dual instances are related, fix $\hat{x}$ satisfying $A\hat{x} = b$ and $(\hat{y}, \hat{s})$ satisfying $A^*\hat{y} + \hat{s} = c$. Letting L denote the nullspace of A, note by (3.2) that up to an additive constant in the objective functional (specifically, the constant $-\langle b, \hat{y} \rangle$), the primal instance is precisely

$$\begin{aligned} \min \quad & \langle \hat{s}, x \rangle \\ \text{s.t.} \quad & x \in L + \hat{x}, \\ & x \in K. \end{aligned} \quad (3.3)$$

On the other hand, making the standard (simplifying) assumption that A is surjective, and hence A^* is injective, if (y, s) is dual feasible, then y is uniquely determined by s. Consequently, the dual instance is equivalent to

$$\begin{aligned} \min \quad & \langle \hat{x}, s \rangle \\ \text{s.t.} \quad & s \in L^\perp + \hat{s}, \\ & s \in K^*. \end{aligned} \quad (3.4)$$

(In fact, they are equivalent even if A is not surjective.) Geometrical relations between the primal and dual instances are apparent from (3.3) and (3.4). Moreover, since $(K^*)^* = K$ (as follows from Corollary 3.2.2 and the closedness of K), (3.3) and (3.4) make it geometrically clear that the dual of the dual instance is the primal instance, a fact that is also readily established algebraically.

In the literature, CP is often introduced without reference to an inner product. The objective functional $\langle c, x \rangle$ is expressed $c^*(x)$, where c^* is an element of the dual space of the primal space $\mathbb{R}^n$, i.e., an element of the space of all continuous linear functionals on the primal space. For us, the inner product naturally identifies the primal space with the dual space; for each element c^* in the dual space there exists c in the primal space such that $c^*(x) = \langle c, x \rangle$ for all x. The main advantage of relying on an inner product is that it makes apparent geometrical relations between a primal instance and its dual instance. The main disadvantage is that the identification of the primal space with the dual space changes when the inner product is changed, forcing one to use notation that depends on the inner product.

3.1. Conic Programming

However, by this point the reader is accustomed to changing local inner products, so the advantages far outweigh the disadvantages.

Let us be precise as to how the primal and dual instances change when the inner product is changed. Assume the inner product on $\mathbb{R}^n$ is changed from $\langle \ , \ \rangle$ to $\langle \ , \ \rangle_H$, where H is pd w.r.t. $\langle \ , \ \rangle$. (Recall $\langle u, v \rangle_H := \langle u, Hv \rangle$.) The primal instance can then be expressed as

$$\begin{aligned} \min \quad & \langle c_H, x \rangle_H \\ \text{s.t.} \quad & Ax = b, \\ & x \in K, \end{aligned}$$

where $c_H := H^{-1}c$. Assuming the inner product on $\mathbb{R}^m$, the range space of A, is left unchanged, the dual instance is

$$\begin{aligned} \max \quad & \langle b, y \rangle \\ \text{s.t.} \quad & A_H^* y + s_H = c_H, \\ & s_H \in K_H^*, \end{aligned}$$

where $A_H^* = H^{-1} A^*$ is the new adjoint of A, $K_H^* = H^{-1} K^*$ is the new dual cone of K, and s_H is the vector of slack variables. A point (y, s) is dual feasible before the change of inner product iff the point $(y, H^{-1}s)$ is dual feasible after the change.

We close this section with an overview of the remainder of the chapter.

In LP, one has *strong duality*; if either val or val* is finite, then val = val*. Strong duality can fail for CP instances, even for simple SDP instances over the 2×2 symmetric matrices. However, under conditions which typically must be satisfied for the application of ipm's, strong duality is present. For example, it is present if the primal and dual instances are feasible and at least one of them remains feasible whenever its right-hand-side vector (b or c, respectively) is perturbed by a small amount, as is the case, say, if A is surjective and the primal instance has a feasible point in the interior of K. We prove this result, as well as other basic results in the classical duality theory, in §3.2.

We study conjugate functionals in §3.3. Given a barrier functional f whose domain is the interior of K, we show (among other things) that the conjugate functional is a barrier functional whose domain is the interior of the dual cone K^*. The connections between ipm's and duality theory hinge critically on this fact.

In §3.4 we establish the fundamental relations between the central path for the primal instance and the central path for the dual instance. We show that in following one path, the other is virtually generated as a by-product.

In §3.5 we present the theory of self-scaled (or symmetric) cones, a topic first developed in the ipm literature by Nesterov and Todd [16], [17]. The pronounced structure of these cones allows for the development of symmetric primal-dual ipm's (algorithms in which the roles of the primal and dual instances are mirror images). The most important cones—both practically and theoretically—are self-scaled cones. The nonnegative orthant, the cone of psd matrices, and the second-order cone are all self-scaled.

In designing an iterative algorithm (an algorithm that generates a sequence of points as do ipm's), the choice of direction for moving from one point to the next is crucial. In §3.6 we discuss the Nesterov–Todd directions, perhaps the most prevalent directions appearing in the design of primal-dual ipm's.

We present and analyze two types of primal-dual ipm's. In §3.7 we consider path-following methods, algorithms which stay near the central path, much like the barrier method in §2.4. In §3.8 we consider a potential-reduction method. Whereas the theory of path-following methods requires the algorithms to stay near the central path, the theory of potential-reduction methods does not, thus suggesting that potential-reduction methods are more robust. The analysis of the progress for a potential-reduction method depends on showing that a certain function—an appropriately chosen potential function—decreases by a constant amount with each iteration. The decrease can be established regardless of where the iterates lie; they need not lie near the central path.

3.2 Classical Duality Theory

In this section we develop the basic duality theory of CP. Although the duality theory predates ipm's by decades, we include its development because the ipm literature often assumes the reader to be familar with the central results. Understanding the proofs in this section is by no means essential for understanding the rest of the chapter. Perhaps the main purpose of this section is to serve as motivation for subsequent sections.

We remarked that for CP, the strong duality relation can fail to hold between a primal instance

$$\begin{aligned} \min \quad & \langle c, x \rangle \\ \text{s.t.} \quad & Ax = b, \\ & x \in K \end{aligned}$$

and its dual instance

$$\begin{aligned} \max \quad & \langle b, y \rangle \\ \text{s.t.} \quad & A^*y + s = c, \\ & s \in K^*. \end{aligned}$$

For an example where strong duality fails, let K be the second-order cone in $\mathbb{R}^3$:

$$K := \{x \in \mathbb{R}^3 : x_1^2 + x_2^2 \leq x_3^2 \text{ and } x_3 \geq 0\}.$$

With regards to the dot product, it is an instructive exercise to show $K = K^*$; that is, K is *self-dual*. Consequently, for the primal instance

$$\begin{aligned} \min \quad & -x_1 \\ \text{s.t.} \quad & x_2 + x_3 = 0, \\ & x \in K, \end{aligned}$$

the dual instance, involving a single variable y, is

$$\begin{aligned} \max \quad & 0y \\ \text{s.t.} \quad & -(1, y, y) \in K. \end{aligned}$$

Rewriting $x_1^2 + x_2^2 \leq x_3^2$ as $x_1^2 \leq (x_2 + x_3)(x_3 - x_2)$, it is readily seen that val $= 0$. On the other hand, to be feasible for the dual, y must satisfy $1 + y^2 \leq y^2$, obviously impossible. Thus val* $= -\infty$ and hence val $\neq$ val*, even though one of the optimal objective values is finite.

3.2. Classical Duality Theory

A slightly more elaborate example shows strong duality can fail even when *both* of the optimal objective values are finite. Consider the primal instance

$$\begin{aligned} \min \quad & -x_1 \\ \text{s.t.} \quad & x_1 + x_4 = 1, \\ & x_2 + x_3 = 0, \\ & x \in K, \end{aligned}$$

where

$$K := \{x \in \mathbb{R}^4 : x_1^2 + x_2^2 \leq x_3^2 \text{ and } x_3, x_4 \geq 0\}.$$

Once again, K is self-dual. Consequently, the dual instance is

$$\begin{aligned} \max \quad & y_1 \\ \text{s.t.} \quad & -(1 + y_1, y_2, y_2, y_1) \in K. \end{aligned}$$

As before, val $= 0$. On the other hand, $y = (-1, 0)$ is dual optimal as all dual feasible points satisfy $y_1 = -1$. Hence val$^* = -1$.

A standard approach to proving strong duality for LP is via the Farkas lemma. The Farkas lemma is a "theorem of exclusive alternatives," stating that exactly one of the following two systems is feasible:

$$\begin{array}{ll} Ax = b, & A^T y \leq 0, \\ x \geq 0, & b^T y > 0. \end{array}$$

The analogue for CP would be that exactly one of the following two systems is feasible:

$$\begin{array}{ll} Ax = b, & -A^* y \in K^*, \\ x \in K, & \langle b, y \rangle > 0. \end{array}$$

However, just as strong duality can fail for CP instances, so can this analogue of the Farkas lemma, as the reader can readily verify from the following example in which K is the second-order cone in $\mathbb{R}^3$:

$$\begin{array}{ll} x_1 = 1, & \\ x_2 + x_3 = 0, & -(y_1, y_2, y_2) \in K, \\ x \in K, & y_1 > 0. \end{array} \tag{3.5}$$

Neither system is feasible.

Although strong duality can fail in CP, something only slightly weaker always holds, something that is called "asymptotic strong duality." Similarly, although the strong version of the Farkas lemma can fail in CP, something only slightly weaker never fails. Moreover, in virtually the same manner one deduces strong duality from the Farkas lemma in LP, one can prove asymptotic strong duality from an asymptotic version of the Farkas lemma for general CP. We do precisely this.

The heart of CP duality theory is the Hahn–Banach theorem. From this single theorem the theory is built. The Hahn–Banach theorem has many versions, most pertaining to very general vector spaces, e.g., arbitrary topological vector spaces, including infinite-dimensional ones. Likewise, CP duality theory can be developed quite generally (cf. [1]

for an extremely general development). However, by restricting to $\mathbb{R}^n$, the proofs are eased considerably. Hence, just as we are developing ipm theory in $\mathbb{R}^n$, we develop duality theory in finite dimensions.

Theorem 3.2.1 (a geometric "Hahn–Banach theorem"). *Assume S is a nonempty, closed, convex subset of $\mathbb{R}^n$. If $x \in \mathbb{R}^n$ but $x \notin S$, then there exists $d \in \mathbb{R}^n$ such that*

$$\sup\{\langle d, s \rangle : s \in S\} < \langle d, x \rangle.$$

Proof. Using the fact that in $\mathbb{R}^n$, bounded sequences have convergent subsequences, it is not difficult to prove that S has a closest point to x, i.e., a point $s' \in S$ satisfying

$$\|s' - x\| = \inf\{\|s - x\| : s \in S\},$$

where $\| \ \|$ is the norm induced by $\langle \ , \ \rangle$.

We claim
$$\langle x - s', s - s' \rangle \leq 0 \quad \text{for all } s \in S. \tag{3.6}$$

Indeed, assume otherwise, letting $s \in S$ violate the inequality. For $0 \leq t \leq 1$, define $s(t) := s' + t(s - s')$. Since $s(t) \in S$ and

$$\|s(t) - x\|^2 = \|s' - x\|^2 + t^2 \|s - s'\|^2 - 2t \langle x - s', s - s' \rangle,$$

we conclude that for sufficiently small $t > 0$, $s(t)$ is a point in S strictly closer to x than is s', a contradiction.

Let $d := x - s'$. By (3.6),

$$\sup\{\langle d, s \rangle : s \in S\} \leq \langle d, s' \rangle \quad \text{for all } s \in S.$$

However,
$$\langle d, s' \rangle = \langle d, x \rangle - \|d\|^2 < \langle d, x \rangle,$$

completing the proof. $\square$

Corollary 3.2.2. *Assume K is a nonempty, closed, convex cone in $\mathbb{R}^n$. If $x \in \mathbb{R}^n$ but $x \notin K$, there exists $d \in \mathbb{R}^n$ such that*

$$\max\{\langle d, s \rangle : s \in K\} = 0 < \langle d, x \rangle.$$

The Farkas lemma for LP extends to CP if one relaxes the notion of feasibility. In LP, an instance is either feasible, or is infeasible and remains infeasible if the right-hand-side vector b is perturbed to any vector $b + \Delta b$ for which $\|\Delta b\|$ is sufficiently small. This is not true of CP instances in general. For example, the system on the left of (3.5) becomes feasible if the right-hand-side coordinate 0 is replaced by $\epsilon > 0$.

Assuming $\mathbb{R}^m$, like $\mathbb{R}^n$, is endowed with an inner product (and the induced norm), one says the system
$$Ax = b,$$
$$x \in K$$

3.2. Classical Duality Theory

is *asymptotically feasible* if it is feasible or can be made feasible by perturbing b by an arbitrarily small amount, that is, if for each $\epsilon > 0$ there exists Δb satisfying $\|\Delta b\| < \epsilon$, where the system obtained by replacing b with $b + \Delta b$ is feasible.

Theorem 3.2.3 (asymptotic Farkas lemma). *Either the system*
$$Ax = b,$$
$$x \in K$$

is asymptotically feasible or the system
$$-A^*y \in K^*,$$
$$\langle b, y \rangle > 0$$

is feasible, but not both.

Proof. Fixing A, consider the cone consisting of right-hand-side vectors for which the first system is feasible:
$$K(A) := \{b \in \mathbb{R}^m : \exists x \in K \text{ s.t. } Ax = b\}.$$

Clearly, $b \in \overline{K(A)}$ (the closure of $K(A)$) iff the system with right-hand-side vector b is asymptotically feasible. Consequently, invoking Corollary 3.2.2 for the cone $\overline{K(A)}$, one has the mutually exclusive alternatives that either the system with right-hand-side vector b is asymptotically feasible or there exists $y \in \mathbb{R}^m$ satisfying

$$\langle \check{b}, y \rangle \leq 0 \text{ for all } \check{b} \in \overline{K(A)}, \tag{3.7}$$
$$\langle b, y \rangle > 0.$$

Observe that y satisfies (3.7) iff y satisfies
$$\langle Ax, y \rangle \leq 0 \text{ for all } x \in K,$$
$$\langle b, y \rangle > 0,$$

that is, iff y satisfies
$$\langle x, A^*y \rangle \leq 0 \text{ for all } x \in K, \tag{3.8}$$
$$\langle b, y \rangle > 0.$$

Since (3.8) is the same as
$$-A^*y \in K^*,$$
$$\langle b, y \rangle > 0,$$

the proof of the theorem is complete. □

Now we turn to discussing asymptotic strong duality. The notion of asymptotic feasibility wears a new guise, one pertaining to the optimal objective value. For the primal instance, the relevant quantity is not the optimal objective value for the instance itself but, rather, the infimum of the optimal objective values from instances obtained by perturbing the right-hand-side vector b by an arbitrarily small amount. This quantity is called the *asymptotic optimal value* for the instance. We denote it a-val. Precisely,

$$\text{a-val} := \lim_{\epsilon \downarrow 0} \inf_{\|\Delta b\| < \epsilon} \inf\{\langle c, x \rangle : x \in K \text{ and } Ax = b + \Delta b\}.$$

Note that a-val = ∞ if the primal instance is not asymptotically feasible.

Here is the main theorem relating the optimal objective values of the primal and dual instances.

Theorem 3.2.4 (asymptotic strong duality). *If the primal instance is asymptotically feasible, then* a-val = val*.

Proof. Fix $v \in \mathbb{R}$ and consider the following constraints in the variables $(x, \alpha) \in \mathbb{R}^n \times \mathbb{R}$:

$$\begin{aligned} Ax &= b, \\ \langle c, x \rangle + \alpha &= v, \\ (x, \alpha) &\in K \times \mathbb{R}_+. \end{aligned} \quad (3.9)$$

Endow $\mathbb{R}^n \times \mathbb{R}$ with the inner product

$$\langle (x_1, \alpha_1), (x_2, \alpha_2) \rangle = \langle x_1, x_2 \rangle + \alpha_1 \alpha_2.$$

The dual cone for $K \times \mathbb{R}_+$ is then $K^* \times \mathbb{R}_+$. Likewise, extend the inner product on $\mathbb{R}^m$ to $\mathbb{R}^m \times \mathbb{R}$. The adjoint of the linear operator

$$(x, \alpha) \mapsto (Ax, \langle c, x \rangle + \alpha)$$

is then

$$(y, \beta) \mapsto (A^* y + \beta c, \beta).$$

It is readily seen that the system (3.9) is asymptotically feasible iff a-val $\leq v$. Consequently, by Theorem 3.2.3 we have the exclusive alternatives that either a-val $\leq v$ or the following system in the variables $(y, \beta) \in \mathbb{R}^m \times \mathbb{R}$ is consistent:

$$\begin{aligned} -(A^* y + \beta c, \beta) &\in K^* \times \mathbb{R}_+, \\ \langle b, y \rangle + \beta v &> 0. \end{aligned} \quad (3.10)$$

We now prove a-val $\geq$ val*. Assume otherwise and choose v to satisfy a-val $< v <$ val*. Since a-val $< v$, we know the system (3.10) is inconsistent. Since $v <$ val*, we know there exists dual feasible y satisfying $\langle b, y \rangle > v$. Then $(y, -1)$ satisfies the system (3.10), a contradiction. Hence a-val $\geq$ val*.

Now we prove a-val $\leq$ val*. Assume otherwise and choose v to satisfy val* $< v <$ a-val. Since $v <$ a-val, the system (3.10) is satisfied by some point (y, β). If $\beta = 0$, then y satisfies

$$\begin{aligned} -A^* y &\in K^*, \\ \langle b, y \rangle &> 0, \end{aligned}$$

and hence, by Theorem 3.2.3, the primal instance is not asymptotically feasible, contradicting the assumption of the theorem. Hence, $\beta < 0$. Consequently, $1/\beta$ is defined and thus the point $\bar{y} := \frac{-1}{\beta} y$ satisfies

$$\begin{aligned} c - A^* \bar{y} &\in K^*, \\ \langle b, \bar{y} \rangle &\geq v, \end{aligned}$$

contradicting val* $< v$. Hence, a-val $\leq$ val*, completing the proof. □

3.2. Classical Duality Theory

Not surprisingly, the theorem has a dual analogue which can be proven by showing that the dual of the dual instance is (equivalent to) the primal instance, then invoking the theorem. To that end, define the asymptotic optimal value for the dual instance to be

$$\text{a-val}^* := \lim_{\epsilon \downarrow 0} \sup_{\|\Delta c\| < \epsilon} \sup\{\langle b, y \rangle : y \in \mathbb{R}^m \text{ and } c + \Delta c - A^* y \in K^*\}.$$

Corollary 3.2.5. *If the dual instance is asymptotically feasible, then* a-val* = val.

Proof. The dual instance is equivalent to

$$\begin{aligned} \min \quad & \langle -b, y \rangle \\ \text{s.t.} \quad & A^* y + s = c, \\ & (y, s) \in \mathbb{R}^m \times K^*. \end{aligned} \tag{3.11}$$

Relying on Corollary 3.2.2 and the closedness of K, it is straightforward to prove $K^{**} = K$. Consequently, the dual cone for the cone $\mathbb{R}^m \times K^*$ is $\{0\} \times K$. Since the adjoint of $[A^* \ I] : \mathbb{R}^m \times \mathbb{R}^n \to \mathbb{R}^n$ is

$$x \mapsto (Ax, x),$$

it follows that the dual instance for (3.11) is

$$\begin{aligned} \max \quad & \langle c, x \rangle \\ \text{s.t.} \quad & -(Ax + b, x) \in \{0\} \times K. \end{aligned} \tag{3.12}$$

Clearly, this instance is equivalent to the original primal instance.

Applying Theorem 3.2.4 to the instance (3.11) and its dual instance (3.12) by the equivalences noted above, the corollary is proven. $\square$

Despite CP duality theory having to make special amends for the possible failure of strong duality, the failure is rare in a generic sense as the next theorem indicates.

Let us say that the primal instance is *strongly feasible* if it is feasible and remains feasible for all sufficiently small perturbations of b (i.e., if there exists $\epsilon > 0$ such that when b is replaced by $b + \Delta b$ for any Δb satisfying $\|\Delta b\| < \epsilon$, the resulting constraints are consistent). Similarly, we can speak of strong feasibility of the dual instance.

Theorem 3.2.6. *If either the primal instance or the dual instance is strongly feasible, then* val = val*.

Proof. We first observe that we may assume both instances are asymptotically feasible. For if, say, the primal instance is not asymptotically feasible—hence val = ∞—there exists y satisfying the second system of Theorem 3.2.3. With the dual instance being strongly feasible—in particular, being feasible—by adding arbitrarily large positive multiples of y to any feasible point for the dual instance, we obtain dual feasible points with arbitrarily large objective values. Thus val* = ∞ = val, giving the equality in the theorem. Hence, assume both instances are asymptotically feasible.

We know that both instances being asymptotically feasible implies

$$\text{a-val} = \text{val}^* \leq \text{val} = \text{a-val}^*. \tag{3.13}$$

For definiteness, assume the dual instance to be strongly feasible. (A similar proof applies if the primal instance is strongly feasible.) In light of the relations (3.13), to prove the theorem it suffices to assume a-val < val and show it follows that the dual instance is not strongly feasible, a contradiction.

So assume a-val < val. By the definition of a-val and the assumption that the primal instance is asymptotically feasible, there exist sequences $\{x_i\} \subset K$, $\{\Delta b_i\}$ such that

$$Ax_i = b + \Delta b_i, \quad \Delta b_i \to 0, \quad \text{and} \quad \langle c, x_i \rangle \to \text{a-val}.$$

We claim $\|x_i\| \to \infty$. For otherwise, $\{x_i\}$ has a convergent subsequence; the limit point x is easily proven to satisfy

$$x \in K, \quad Ax = b, \quad \text{and} \quad \langle c, x \rangle = \text{a-val},$$

from which it is immediate that a-val = val, contrary to our assumption a-val < val.

Let Δc be a limit point of the sequence $\{\frac{-1}{\|x_i\|} x_i\}$. Of course $\|\Delta c\| = 1$; in particular, $\Delta c \neq 0$. For fixed $\epsilon > 0$ consider the following CP instance:

$$\begin{aligned} \min \quad & \langle c + \epsilon \Delta c, x \rangle \\ \text{s.t.} \quad & Ax = b, \\ & x \in K. \end{aligned} \qquad (3.14)$$

Note the asymptotic optimal value of this instance is no greater than

$$\begin{aligned} \liminf_i \langle c + \epsilon \Delta c, x_i \rangle &= \text{a-val} + \epsilon \liminf_i \langle \Delta c, x_i \rangle \\ &= \text{a-val} - \epsilon \lim_i \|x_i\| \\ &= -\infty. \end{aligned}$$

Hence, by Theorem 3.2.4, the dual instance of (3.14) is infeasible. The dual instance is

$$\begin{aligned} \max \quad & \langle b, y \rangle \\ \text{s.t.} \quad & (c + \epsilon \Delta c) - A^* y \in K^*. \end{aligned} \qquad (3.15)$$

Since ϵ can be an arbitrarily small positive number, this infeasibility contradicts the assumed strong feasibility of the original dual instance. □

In applying ipm's to solve a primal CP instance, one relies on a barrier functional whose domain is the interior of the cone K. Thus, for the central path to exist, there must be primal feasible points in the interior of K. In that regard, the following corollary shows the potential failure of strong duality to be somewhat irrelevant to the study of ipm's. The corollary implies that if the central path exists for the primal instance, then strong duality holds. The same holds for the dual instance (as is further elucidated in §3.4).

Corollary 3.2.7. *If the linear operator A is surjective and there is a primal feasible point in the interior of K, then val = val*. Similarly, if there exists a dual feasible point (y, s) with s in the interior of K^*, then val = val*.*

Lastly, we briefly discuss the fact that in CP there may not exist optimal solutions even when the optimal objective values are finite—and even when, in addition, strong duality

3.3. The Conjugate Functional

is present. Strictly speaking, our use of "min" for the primal instance and "max" for the dual instance should be replaced by "inf" and "sup," respectively. However, under an assumption of strong feasibility, an optimal solution does indeed exist. Insofar as we are interested in the connections between duality theory and ipm theory, "min" and "max" are entirely appropriate.

For an example of an optimal solution not existing, let K denote the second-order cone in $\mathbb{R}^3$. We leave it to the reader to verify that the primal instance

$$\begin{aligned} \inf \quad & x_2 + x_3 \\ \text{s.t.} \quad & x_1 = 1, \\ & x \in K \end{aligned}$$

satisfies val $= 0$, the dual instance satisfies val$^* = 0$, but the primal instance does not have an optimal solution.

Theorem 3.2.8. *If the dual instance is strongly feasible and the primal instance is feasible, then the primal instance has an optimal solution. Similarly, if the primal instance is strongly feasible and the dual instance is feasible, then the dual instance has an optimal solution.*

Proof. The proof is similar in spirit to that of Theorem 3.2.6.

We consider the case that the dual instance is strongly feasible and the primal instance is feasible. Since the primal instance is feasible, there exists a feasible sequence $\{x_i\}$ satisfying $\langle c, x_i \rangle \to$ val. If the sequence is bounded, it has limit points and those points are easily argued to be optimal for the primal instance. Hence, we may assume the sequence to be unbounded. Then, exactly as in the proof of Theorem 3.2.6, we can perturb the vector c by an arbitrarily small amount to obtain a dual instance (3.15) which is infeasible, contradicting the strong feasibility of the dual instance.

The other case is handled similarly. □

It should be mentioned that strong duality and the Farkas lemma for LP can be seen as special cases of the asymptotic analogues for CP. In the CP theory, the need for asymptotic results has its roots in the cone $K(A)$ appearing in the proof of Theorem 3.2.3. The culprit is the potential nonclosedness of that cone. However, if the cone K is polyhedral (e.g., $K = \mathbb{R}^n_+$, as in LP), then so is $K(A)$, and hence $K(A)$ is closed. With $K(A)$ closed, a strong version of the Farkas lemma is obtained, and from it, strong duality.

3.3 The Conjugate Functional

The ipm literature is full of interesting relations between primal instances

$$\begin{aligned} \min \quad & \langle c, x \rangle \\ \text{s.t.} \quad & Ax = b, \\ & x \in K \end{aligned}$$

and their dual instances

$$\begin{aligned} \max \quad & \langle b, y \rangle \\ \text{s.t.} \quad & A^*y + s = c, \\ & s \in K^*. \end{aligned}$$

For these relations, one assumes K°—the interior of K—is the domain of a barrier functional.

To get a sense of how a barrier functional $f : K^\circ \to \mathbb{R}$ joins with duality theory, consider the (negative) gradient map $x \mapsto -g(x)$. For example, in LP with the dot product, this is the map $(x_1, \ldots, x_n) \mapsto (\frac{1}{x_1}, \ldots, \frac{1}{x_n})$. We claim the map $x \mapsto -g(x)$ takes K° into K^*. Indeed, by Theorem 2.3.3, whenever $x, x' \in K^\circ$,
$$\langle g(x), x' - x \rangle < \vartheta_f.$$
Applying this with tx' in place of x' and letting $t \to \infty$, one deduces
$$\langle -g(x), x' \rangle \geq 0 \quad \text{for all } x' \in K,$$
that is, $-g(x) \in K^*$ as claimed.

The gradient map is injective, i.e., one-to-one. For if x and x' satisfy $g(x) = g(x')$, then both x and x' minimize the strictly convex functional $\bar{x} \mapsto -\langle g(x), \bar{x} \rangle + f(\bar{x})$, forcing $x = x'$.

We shall see (Theorem 3.3.1) that not only does the gradient map $x \mapsto -g(x)$ carry K° injectively into K^*, it carries K° onto $(K^*)^\circ$, the interior of K^*. The map is a bijection between K° and $(K^*)^\circ$. Furthermore, we shall see that the resulting inverse map is itself the (negative) gradient map for a barrier functional, namely, for the *conjugate functional*[1] of f:
$$f^*(s) := -\inf_{x \in D_f} (\langle x, s \rangle + f(x)).$$
The conjugate functional played a prominent role in optimization long before ipm research blossomed, long before the notion of a self-concordant functional took form (cf. [19]).

Later in this chapter when local inner products become useful, it will be important to keep in mind that the definition of the conjugate functional, like the definition of the dual cone, depends on the underlying inner product. When the inner product changes, so does the conjugate functional.

The definition of the conjugate functional f^* applies to each functional f, not just barrier functionals. Regardless of f, the conjugate functional is convex, for it is the supremum of the convex functionals—in fact, linear functionals—
$$s \mapsto -\langle x, s \rangle - f(x).$$
We define the domain D_{f^*} to be the set consisting of all s for which $f^*(s)$ is finite.

Throughout this section we assume f to satisfy only the following properties unless otherwise stated: $f \in C^2$, D_f is open and convex, $H(x)$ is pd for all $x \in D_f$.

Recall $\mathcal{SC}$ denotes the set of (strongly nondegenerate) self-concordant functionals and $\mathcal{SCB}$ denotes the subset of $\mathcal{SC}$ composed of barrier functionals.

It is suggested that on a first pass, the reader skip proofs in this section.

Theorem 3.3.1. *If $f \in \mathcal{SC}$, then $f^* \in \mathcal{SC}$. If $f \in \mathcal{SCB}$ and $D_f = K^\circ$, where K is a cone, then $f^* \in \mathcal{SCB}$ and*
$$D_{f^*} = \{-g(x) : x \in K^\circ\} = (K^*)^\circ.$$

---
[1] This differs slightly from the standard definition of the conjugate functional: $s \mapsto \sup_{x \in D_f} \langle x, s \rangle - f(x)$. The only real effects of the difference are to reduce the number of minus signs appearing in our exposition and to make the domain of the conjugate functional be the dual cone rather than the polar cone.

3.3. The Conjugate Functional

Moreover, $\vartheta_{f^*} \leq (4\vartheta_f + 1)^2$. Furthermore, if f is logarithmically homogeneous, then so is f^* and $\vartheta_{f^*} = \vartheta_f$.

Toward proving the theorem, we present two propositions and a theorem.

Proposition 3.3.2. *If $f \in \mathcal{SCB}$ and $D_f = K^\circ$, where K is a cone, then $D_{f^*} = (K^*)^\circ$.*

This proposition will be proven using the following proposition, which figures prominently in proofs throughout this section.

Proposition 3.3.3. *The gradient map $g : D_f \to \mathbb{R}^n$ is injective. If $f \in SC$, then*

$$D_{f^*} = \{-g(x) : x \in D_f\} \tag{3.16}$$

and D_{f^*} is open.

Proof. If $-g(x_1) = s = -g(x_2)$, then x_1 and x_2 minimize the strictly convex functional

$$x \mapsto \langle x, s \rangle + f(x). \tag{3.17}$$

Hence, $x_1 = x_2$ and thus g is injective. Moreover, because the functional (3.17) then has a minimizer, $s \in D_{f^*}$. Hence,

$$\{-g(x) : x \in D_f\} \subseteq D_{f^*}. \tag{3.18}$$

Assume $s \in D_{f^*}$ and $f \in \mathcal{SC}$. Since D_{f^*} consists of s for which $f^*(s)$ is finite, the self-concordant functional (3.17) is bounded below. By Theorem 2.2.8, it has a minimizer x. Clearly, $-g(x) = s$. Hence,

$$D_{f^*} \subseteq \{-g(x) : x \in D_f\},$$

and so with (3.18), we have (3.16).

The openness of D_{f^*} now follows from the differential of g—i.e., the Hessian—being surjective. Alternatively, openness follows from Proposition 2.2.10. □

Proof of Proposition **3.3.2.** In the opening paragraphs of this section we saw $\{-g(x) : x \in K^\circ\} \subseteq K^*$. Hence, by Proposition 3.3.3, $D_{f^*} \subseteq (K^*)^\circ$. Thus, again by Proposition 3.3.3, to complete the proof it suffices to show that if $s \in (K^*)^\circ$, then there exists $x \in K^\circ$ such that $-g(x) = s$; that is, it suffices to show the self-concordant functional

$$\bar{f}(x) := \langle x, s \rangle + f(x)$$

has a minimizer.

Since $s \in (K^*)^\circ$, there exists $\alpha > 0$ such that

$$\langle x, s \rangle \geq \alpha \|x\| \text{ for all } x \in K.$$

In particular, if $x \in K^\circ$, then

$$\langle x, s \rangle + \langle g(x), x \rangle = \langle x, s \rangle - \langle g(x), 0 - x \rangle$$
$$\geq \alpha \|x\| - \vartheta_f, \tag{3.19}$$

the inequality by Theorem 2.3.3. Observe that if $\|x\| \geq \vartheta_f/\alpha$, then (3.19) implies the functional $\bar{f}$ to be strictly increasing as one moves outward along the ray $\{x + tx : t \geq 0\}$. Hence,
$$\inf\{\bar{f}(x) : x \in K^\circ\} = \inf\{\bar{f}(x) : x \in K^\circ \text{ and } \|x\| \leq \vartheta_f/\alpha\}. \tag{3.20}$$

For the self-concordant functional $\bar{f}$, Theorem 2.2.9 shows $\bar{f}(x_i) \to \infty$ if a sequence $\{x_i\}$ converges to a point in the boundary of K. Consequently, using (3.20), it is easily argued by compactness that $\bar{f}$ has a minimizer. □

The identities stated in the next theorem are absolutely fundamental in developing primal-dual ipm's.

Theorem 3.3.4. *Assume $f \in SC$. Then $f^* \in C^2$. Moreover, if x and s satisfy $s = -g(x)$, then*
$$-g^*(s) = x \quad \text{and} \quad H^*(s) = H(x)^{-1},$$
where g^* and H^* denote the gradient and Hessian of f^*.

Proof. The proof goes somewhat outside this book, relying on differentials and the inverse function theorem.

By Proposition 3.3.3, the map $-g : D_f \to D_{f^*}$ is invertible. For each $s \in D_{f^*}$, let $x(s)$ denote the point in D_f satisfying $-g(x(s)) = s$. Since $g \in C^1$ and the first-differentials of g are invertible (i.e., the Hessians $H(x)$ are invertible), the inverse function theorem implies $s \mapsto x(s)$ is a C^1 map.

Differentiating both sides of the identity $-g(x(s)) = s$ w.r.t. s by making use of the chain rule, we have
$$-H(x(s))D_s x(s) = I,$$
that is,
$$D_s x(s) = -H(x(s))^{-1}. \tag{3.21}$$

Let $[D_s x(s)]^T$ denote the adjoint of $D_s x(s) : \mathbb{R}^n \to \mathbb{R}^n$. (We use "$T$" because "$*$" is in use.)

Clearly,
$$f^*(s) = -\langle x(s), s \rangle - f(x(s)).$$
Since the map $s \mapsto x(s)$ is in C^1, so is the functional f^*. Differentiating both sides w.r.t. s, making use of the product rule and the chain rule, we have
$$g^*(s) = -x(s) - [D_s x(s)]^T s - [D_s x(s)]^T g(x(s))$$
$$= -x(s), \tag{3.22}$$

the last equality because $s = -g(x(s))$. Thus, whenever x and s satisfy $-g(x) = s$ we have $-g^*(s) = x$. Moreover, since $s \mapsto x(s)$ is a C^1 mapping, (3.22) trivially shows the same to be true of g^*. Hence, $f^* \in C^2$.

Finally, note (3.22) and (3.21) imply
$$H^*(s) = -D_s x(s) = H(x(s))^{-1};$$
that is, if $-g(x) = s$, then $H^*(s) = H(x)^{-1}$. □

3.3. The Conjugate Functional

Proof of Theorem 3.3.1. We use a superscript "$*$" to distinguish between objects associated with f^* and those associated with f. For example, $\|\ \|_s^*$ denotes the local norm associated with f^*.

Assume $f \in \mathcal{SC}$. Assume x and s satisfy $-g(x) = s$. By Theorem 3.3.4, for all w_1, w_2,

$$\begin{aligned}\langle w_1, w_2\rangle_s^* &= \langle w_1, H^*(s)w_2\rangle \\ &= \langle w_1, H(x)^{-1}w_2\rangle \\ &= \langle H(x)^{-1}w_1, H(x)^{-1}w_2\rangle_x,\end{aligned}$$

an identity we use freely in the remainder of the proof. Likewise for the resulting relation

$$\|w\|_s^* = \|H(x)^{-1}w\|_x.$$

We now prove $f^* \in \mathcal{SC}$. By Theorem 2.5.2, it suffices to show that if $s \in D_{f^*}$, then

$$B_s^*\left(s, \tfrac{1}{4}\right) \subseteq D_{f^*} \tag{3.23}$$

and

$$\limsup_{\Delta s \to 0} \frac{\|I - H_s^*(s + \Delta s)\|_s^*}{\|\Delta s\|_s^*} \leq 2. \tag{3.24}$$

Assume Δs is a vector satisfying $r := \|\Delta s\|_s^* < \tfrac{1}{4}$, that is, $\|v\|_x < \tfrac{1}{4}$, where $v := -H(x)^{-1}\Delta s$. Noting $\frac{3r^2}{(1-r)^3} \leq 9r^2$, Proposition 2.2.10 shows there exists $u \in B_x(v, 9\|v\|_x^2)$ such that

$$g_x(x + u) = g_x(x) + v.$$

Note that

$$\begin{aligned}-g(x + u) &= -H(x)g_x(x + u) \\ &= -H(x)(g_x(x) + v) \\ &= -g(x) - H(x)v \\ &= s + \Delta s.\end{aligned}$$

Consequently, by Proposition 3.3.3, we have $s + \Delta s \in D_{f^*}$, establishing (3.23). Moreover,

$$\|u\|_x \leq \|v\|_x + 9\|v\|_x^2 = \|\Delta s\|_s^* + 9(\|\Delta s\|_s^*)^2. \tag{3.25}$$

Toward establishing (3.24), observe

$$\begin{aligned}\|I - H_s^*(s + \Delta s)\|_s^* &= \max_w \frac{|\langle w, [I - H_s^*(s + \Delta s)]w\rangle_s^*|}{(\|w\|_s^*)^2} \\ &= \max_w \frac{|\langle H(x)w, [I - H_s^*(s + \Delta s)]H(x)w\rangle_s^*|}{(\|H(x)w\|_s^*)^2} \\ &= \max_w \frac{|\langle w, H(x)^{-1}[I - H_s^*(s + \Delta s)]H(x)w\rangle_x|}{(\|w\|_x)^2} \\ &= \max_w \frac{|\langle w, [I - H(x)^{-1}H_s^*(s + \Delta s)H(x)]w\rangle_x|}{(\|w\|_x)^2} \\ &= \|I - H(x)^{-1}H_s^*(s + \Delta s)H(x)\|_x. \tag{3.26}\end{aligned}$$

However,
$$H_s^*(s + \Delta s) = H^*(s)^{-1} H^*(s + \Delta s) = H(x) H(x + u)^{-1},$$
implying
$$H(x)^{-1} H_s^*(s + \Delta s) H(x) = H(x + u)^{-1} H(x) = H_x(x + u)^{-1}. \tag{3.27}$$
By (3.26), (3.27), (3.25), and self-concordance of f we have
$$\|I - H_s^*(s + \Delta s)\|_s^* = \|I - H_x(x + u)^{-1}\|_x$$
$$\leq \frac{1}{(1 - \|u\|_x)^2} - 1$$
$$\leq \frac{1}{(1 - \|\Delta s\|_s^* - 9(\|\Delta s\|_s^*)^2)^2} - 1,$$
from which (3.24) is immediate. Hence, $f^* \in \mathcal{SC}$.

Assume $f \in \mathcal{SCB}$ and $D_f = K^\circ$, where K is a cone. Proposition 3.3.2 shows $D_{f^*} = (K^*)^\circ$. To see that f^* is a barrier functional, first observe
$$\|g_s^*(s)\|_s^* = \|H^*(s)^{-1} g^*(s)\|_s^*$$
$$= \|H(x) x\|_s^*$$
$$= \|x\|_x. \tag{3.28}$$
If f is logarithmically homogeneous, Theorem 2.3.9 shows $x = -g_x(x)$, and hence (3.28) gives
$$\vartheta_{f^*} := \sup_{s \in D_{f^*}} (\|g_s^*(s)\|_s^*)^2$$
$$= \sup_{x \in D_f} \|x\|_x^2$$
$$= \sup_{x \in D_f} \|g_x(x)\|_x^2$$
$$= \vartheta_f.$$
Even if f is not logarithmically homogeneous, we have $\langle g(x), 0 - x \rangle \geq 0$ since $-g(x) \in K^*$. Consequently, because $0 \in K$, Theorem 2.3.4 implies
$$\|x\|_x \leq 4\vartheta_f + 1.$$
Hence, by (3.28), $\vartheta_{f^*} \leq (4\vartheta_f + 1)^2$.

Finally, if f is logarithmically homogeneous,
$$g(tx) = \tfrac{1}{t} g(x) = -\tfrac{1}{t} s,$$
and hence, by Theorem 3.3.4,
$$g^* \left(\tfrac{1}{t} s \right) = -tx = t g^*(s);$$
that is, g^* is homogeneous of degree -1. Thus, by Theorem 2.3.9, f^* is logarithmically homogeneous. □

We remark that using the first statement in Theorem 3.3.1 together with the identity for D_{f^*} in Proposition 3.3.3 and the relation $-g^*(s) = x$ in Theorem 3.3.4, it is readily proven for $f \in \mathcal{SC}$ that $f = (f^*)^*$, just as one wants for a perfectly symmetric duality theory.

3.4 Duality of the Central Paths

Let f denote a barrier functional whose domain is the interior of a cone K, i.e., $D_f = K^\circ$. We know from Theorem 3.3.1 that f^*, the conjugate functional, is a barrier functional whose domain is $(K^*)^\circ$, the interior of the dual cone.

Recall that the central path $\{x_\eta : \eta > 0\}$ for the primal instance

$$\begin{aligned} \min \quad & \langle c, x \rangle \\ \text{s.t.} \quad & Ax = b, \\ & x \in K \end{aligned}$$

consists of the points x_η solving the linearly constrained optimization problems

$$\begin{aligned} \min_x \quad & \eta \langle c, x \rangle + f(x) \\ \text{s.t.} \quad & Ax = b. \end{aligned}$$

The central path $\{(y_\eta, s_\eta) : \eta > 0\}$ for the dual instance

$$\begin{aligned} \max \quad & \langle b, y \rangle \\ \text{s.t.} \quad & A^*y + s = c, \\ & s \in K^* \end{aligned}$$

consists of the points (y_η, s_η) solving the problems

$$\begin{aligned} \max_{y,s} \quad & \eta \langle b, y \rangle - f^*(s) \\ \text{s.t.} \quad & A^*y + s = c. \end{aligned}$$

We shall see that under the assumption of logarithmic homogeneity, these central paths are strongly related, each arising from the optimality conditions for the other. For this, it is convenient to adopt the geometrical viewpoint described in §3.1.

Fixing $\hat{x}$ satisfying $A\hat{x} = b$ and $(\hat{y}, \hat{s})$ satisfying $A^*\hat{y} + \hat{s} = c$, recall that the primal instance is identical (up to an additive constant in the objective functional) to

$$\begin{aligned} \min \quad & \langle \hat{s}, x \rangle \\ \text{s.t.} \quad & x \in L + \hat{x}, \\ & x \in K, \end{aligned}$$

where L is the nullspace of A. The dual instance is equivalent to

$$\begin{aligned} \min \quad & \langle \hat{x}, s \rangle \\ \text{s.t.} \quad & s \in L^\perp + \hat{s}, \\ & s \in K^*. \end{aligned}$$

Maintaining our simplifying assumption that A is surjective, and hence A^* is injective, y is uniquely determined by s from $A^*y + s = c$.

The primal central path consists of the points x_η solving the optimization problems

$$\begin{aligned} \min_x \quad & \eta \langle \hat{s}, x \rangle + f(x) \\ \text{s.t.} \quad & x \in L + \hat{x}, \end{aligned} \tag{3.29}$$

whereas the dual central path consists of the points s_η solving the problems

$$\begin{aligned} \min_s \quad & \eta \langle \hat{x}, s \rangle + f^*(s) \\ \text{s.t.} \quad & s \in L^\perp + \hat{s}. \end{aligned} \qquad (3.30)$$

Necessary (and sufficient) conditions for x_η to solve (3.29) are that

$$x_\eta \in L + \hat{x} \quad \text{and} \quad \eta \hat{s} + g(x_\eta) \in L^\perp. \qquad (3.31)$$

Hence, defining

$$\bar{s}_\eta := -\tfrac{1}{\eta} g(x_\eta),$$

by the latter condition we have

$$\bar{s}_\eta \in L^\perp + \hat{s}. \qquad (3.32)$$

Moreover, since Theorem 3.3.1 shows $-g(x_\eta) \in (K^*)^\circ$, and since K is a cone, we have $\bar{s}_\eta \in (K^*)^\circ$. Thus, $\bar{s}_\eta$ is dual feasible.

Assuming logarithmic homogeneity, we claim $\bar{s}_\eta = s_\eta$; that is, the vector $\bar{s}_\eta$ arising from the optimality conditions for x_η being on the primal central path is itself on the dual central path (moreover, with the same parameter value η). For under logarithmic homogeneity we have

$$\begin{aligned} g^*(\bar{s}_\eta) &= g^*(-\tfrac{1}{\eta} g(x_\eta)) \\ &= \eta g^*(-g(x_\eta)) \\ &= -\eta x_\eta, \end{aligned}$$

the last equality because $-g^*$ is the inverse of $-g$ by Theorem 3.3.4 (regardless of whether logarithmic homogeneity holds). Hence, by the first of the necessary conditions (3.31),

$$\eta \hat{x} + g^*(\bar{s}_\eta) \in L.$$

Together with (3.32) we now have that $\bar{s}_\eta$ satisfies the sufficient (and necessary) conditions for s_η to be the unique optimal solution of the strictly convex problem (3.30):

$$s_\eta \in L^\perp + \hat{s} \quad \text{and} \quad \eta \hat{x} + g^*(s_\eta) \in L. \qquad (3.33)$$

Consequently, $\bar{s}_\eta = s_\eta$ as claimed.

Having proved $s_\eta = -\tfrac{1}{\eta} g(x_\eta)$, it is easy to establish $x_\eta = -\tfrac{1}{\eta} g(s_\eta)$, assuming logarithmic homogeneity; indeed, from the fact that $-g^*$ and $-g$ are inverses,

$$\begin{aligned} -\tfrac{1}{\eta} g^*(s_\eta) &= -\tfrac{1}{\eta} g^*\left(-\tfrac{1}{\eta} g(x_\eta)\right) \\ &= -g^*(-g(x_\eta)) \\ &= x_\eta. \end{aligned}$$

Alternatively, just as we relied on the necessity of the conditions (3.31) to prove $s_\eta = -\tfrac{1}{\eta} g(x_\eta)$ by establishing that $\bar{s}_\eta := -\tfrac{1}{\eta} g(x_\eta)$ satisfies the sufficient conditions (3.33), we could rely on the necessity of the conditions (3.33) to prove $x_\eta = -\tfrac{1}{\eta} g^*(s_\eta)$ by establishing that $\bar{x}_\eta := -\frac{1}{\eta} g^*(s_\eta)$ satisfies the sufficient conditions (3.31). The roles of x_η and s_η are entirely symmetric.

An important consequence of the relations $s_\eta = -\frac{1}{\eta} g(x_\eta)$ and $x_\eta = -\frac{1}{\eta} g^*(s_\eta)$ is that by following one of the central paths, one can generate the other as a by-product. The dual instance can be solved by solving the primal instance and vice versa.

We claim that regardless of whether logarithmic homogeneity holds, the dual points $\bar{s}_\eta$ (primal points $\bar{x}_\eta$) tend to optimality as $\eta \to \infty$. For we have already proven that $\bar{s}_\eta$ is dual feasible; that did not rely on logarithmic homogeneity. Moreover, letting $\bar{y}_\eta$ denote the vector satisfying $A^* \bar{y}_\eta + \bar{s}_\eta = c$, we have

$$\begin{aligned}
\langle c, x_\eta \rangle - \langle b, \bar{y}_\eta \rangle &= \langle x_\eta, \bar{s}_\eta \rangle \\
&= \tfrac{1}{\eta} \langle x_\eta, -g(x_\eta) \rangle \\
&= \tfrac{1}{\eta} \langle g(x_\eta), 0 - x_\eta \rangle \\
&\leq \vartheta_f / \eta,
\end{aligned}$$

the inequality by Theorem 2.3.3. Consequently, $\bar{s}_\eta$ does indeed tend to optimality as $\eta \to \infty$, and similarly for $\bar{x}_\eta$.

3.5 Self-Scaled (or Symmetric) Cones

3.5.1 Introduction

A self-scaled (or symmetric) cone is a cone K whose interior K° is the domain of a barrier functional f with particularly strong properties. The properties allow for the development of symmetric primal-dual ipm's, i.e., algorithms in which the roles of the primal and dual instances are mirror images. To state the properties, we introduce some notation and a few definitions.

For $z \in K^\circ$, let K_z^* denote the dual cone of K w.r.t. the intrinsic inner product $\langle \, , \, \rangle_z$, that is,

$$K_z^* := \{s : \langle x, s \rangle_z \geq 0 \text{ for all } x \in K\}.$$

We define K to be *intrinsically self-dual* if $K_z^* = K$ for all $z \in K^\circ$.

Strictly speaking, the property of intrinsic self-duality is a property of f, not of K, for K° may be the domain of a second barrier functional whose intrinsic inner products do not satisfy $K_z^* = K$. However, in the relevant literature it is standard to accord the property to K rather than to f, meaning that the structure of K is sufficiently rich to yield a barrier functional with the property. Henceforth, when a cone K is assumed to be intrinsically self-dual, the reader should take f to denote a barrier functional with the property that $K_z^* = K$ for all $z \in D_f = K^\circ$.

Let f_z^* denote the conjugate of f w.r.t. $\langle \, , \, \rangle_z$:

$$f_z^*(s) := - \inf_{x \in K^\circ} (\langle x, s \rangle_z + f(x)).$$

If K is intrinsically self-dual, then f and f_z^* have K° as their domain (by Theorem 3.3.1).

We define f to be *intrinsically self-conjugate* if f is logarithmically homogeneous, if K is intrinsically self-dual, and if for each $z \in K^\circ$ there exists a constant C_z such that

$f_z^* \equiv f + C_z$, i.e., for all $s \in K^\circ$,
$$f_z^*(s) = f(s) + C_z.$$

The theory of ipm's focuses on the gradients and Hessians of barrier functionals rather than on values of the functionals. The relevance of intrinsic self-conjugateness is that for each local inner product $\langle \, , \, \rangle_z$, the gradients g_z and Hessians H_z of f for the primal setting in CP are exactly the same as the gradients g_z^* and Hessians H_z^* of f_z^* for the dual setting.

A cone K is *self-scaled (or symmetric)* if K° is the domain of an intrinsically self-conjugate barrier functional.

In the definition of intrinsic self-conjugateness, the constant C_z is unrestricted. However, it happens that the various conditions of the definition force the constant to equal a specific value, as is shown by the following proposition.

Proposition 3.5.1. *If $f : K^\circ \to \mathbb{R}$ is an intrinsically self-conjugate barrier functional, then for all $z \in K^\circ$,*
$$C_z = -(\vartheta_f + 2f(z)).$$

Proof. Theorem 3.3.4 shows that regardless of the inner product, the conjugate functional satisfies
$$f^*(s) = \langle g^*(s), s \rangle - f(-g^*(s)).$$

Thus, intrinsic self-conjugateness gives
$$\begin{aligned} f(x) &= f_z^*(x) - C_z \\ &= \langle g_z^*(x), x \rangle_z - f(-g_z^*(x)) - C_z \\ &= \langle g_z(x), x \rangle_z - f(-g_z(x)) - C_z. \end{aligned}$$

In particular,
$$f(z) = \langle g_z(z), z \rangle_z - f(-g_z(z)) - C_z.$$

However, since f is logarithmically homogeneous, we know from Theorem 2.3.9 that $\langle g(z), z \rangle = -\vartheta_f$ and $-g_z(z) = z$. Substitution completes the proof. □

The following theorem provides a useful characterization of intrinsic self-conjugacy.

Theorem 3.5.2. *Assume K is a cone and f is a logarithmically homogeneous barrier functional with domain K°. Then f is intrinsically self-conjugate iff for each $z \in K^\circ$, the map $x \mapsto -g_z(x)$ is an involution, i.e., the range of the map (like the domain) is K° and $-g_z(-g_z(x)) = x$ for all $x \in K^\circ$.*

Proof. If f is intrinsically self-conjugate, then $g_z^* \equiv g_z$ and $H_z^* \equiv H_z$. Hence, Theorem 3.3.4 implies the map $x \mapsto -g_z(x)$ is an involution.

Conversely, assume for each $z \in K^\circ$ that $-g_z(-g_z(x)) = x$ for all $x \in K^\circ$. Since Theorem 3.3.4 asserts $-g_z^*(-g_z(x)) = x$ for all $x \in K^\circ$, it follows that $g_z \equiv g_z^*$. But then f differs from f_z^* by at most a constant, i.e., f is intrinsically self-conjugate. □

The theorem provides an easy means to verify that the logarithmic barrier function for the nonnegative orthant is intrinsically self-conjugate. For letting e denote the vector of

3.5. Self-Scaled (or Symmetric) Cones

all ones (hence $\langle \, , \, \rangle_e$ is the dot product), we have

$$g_z(x) = H_e(z)^{-1} g_e(x) = -\left(\frac{z_1^2}{x_1}, \ldots, \frac{z_n^2}{x_n}\right)^T.$$

Consequently, $-g_z(-g_z(x)) = x$.

The same applies for the logarithmic barrier function for the cone of psd matrices. Letting E denote the identity matrix (hence $\langle \, , \, \rangle_E$ is the trace product), we have

$$g_Z(X) = H_E(Z)^{-1} g_E(X) = Z g_E(X) Z = -Z X^{-1} Z.$$

Hence, $-g_Z(-g_Z(X)) = X$.

Similarly, with somewhat more involved calculations one can use the theorem to show that the logarithmic barrier functional for the second-order cone is intrinsically self-conjugate.

The family of all self-scaled cones is limited. Güler [8] made this known to optimizers (see also Faybusovich [5]). Güler realized that self-scaled cones are the same as symmetric cones, objects that have been studied by analysts for decades (cf. [4]). There are five basic symmetric cones, all others being Cartesian products of these:

- the cone of psd symmetric matrices,
- the second-order cone,
- the cone of psd Hermitian matrices,
- the cone of psd Hermitian quaternion matrices,
- a 27-dimensional exceptional cone.

Nesterov and Todd did not know of the equivalence between self-scaled cones and symmetric cones when they wrote their foundational papers [16], [17]. Their principal motivation was to generalize primal-dual methods beyond LP. Although the generalization they achieved was perhaps not as great as we imagined (because we now know there are only five basic self-scaled cones), they succeeded in uncovering the essence of the conic structure needed by primal-dual methods. One can develop the theory of primal-dual methods for each of the five cones individually, but developing it generally has the advantage of making apparent the essential structure.

3.5.2 An Important Remark on Notation

Throughout the remainder of the book we fix $e \in K^\circ$ and let $\langle \, , \, \rangle = \langle \, , \, \rangle_e$; that is, we suppress the subscript e. Likewise, we write g in place of g_e, H in place of H_e, K^* in place of K_e^*, etc. We do this mainly to avoid excessive notation. The choice of $e \in K^\circ$ is arbitrary, i.e., any point in K° will do. Indeed, as the following development will make apparent, in terms of each point's local inner product, the point's geometric view of the cone coincides exactly with the geometric views of the other points in terms of their local inner products; thus the name "symmetric cone."

Throughout the remainder of the book, whenever K is assumed to be self-scaled, the reader should take f to denote an intrinsically self-conjugate barrier functional with domain

K°. When we refer specifically to the nonnegative orthant, the cone of psd matrices, or the second-order cone, the underlying intrinsically self-conjugate barrier functional is assumed to be the logarithmic barrier function.

3.5.3 Scaling Points

Assume K is self-scaled, fix an arbitrary point $e \in K^\circ$, and let $\langle \, , \, \rangle = \langle \, , \, \rangle_e$. For $z \in K^\circ$ we have
$$K = K_z^* = H(z)^{-1} K^* = H(z)^{-1} K,$$
the second identity being a simple consequence of
$$\langle u, v \rangle_z = \langle u, H(z) v \rangle.$$
Consequently, $H(z)$ is a linear operator carrying K bijectively onto itself, i.e., $H(z)$ is a linear automorphism of K, a "scaling" of K.

The next theorem shows the set of linear automorphisms $\{H(z) : z \in K^\circ\}$ forms a complete set of scalings in the sense that for each $x \in K^\circ$,
$$\{H(z)x : z \in K^\circ\} = K^\circ;$$
that is, for each point in K° there exists some—in fact, a unique—automorphism $H(z)$ carrying x to that point.

Theorem 3.5.3. *If K is self-scaled, then $H(z)$ is a linear automorphism of K for each $z \in K^\circ$. Moreover, if $x, s \in K^\circ$, then there exists a unique $w \in K^\circ$ satisfying*
$$H(w)x = s.$$

(Consequently, if $H(w_1) = H(w_2)$, then $w_1 = w_2$.)

When $H(w)x = s$, the point w is referred to as the *scaling point* (w.r.t the local inner product at e) for the ordered pair x, s. That it is unique proves to be very important in the theory.

If w is the scaling point for the ordered pair x, s, then $-g(w)$ is the scaling point for the pair s, x. In fact,
$$H(-g(w)) = H^*(-g(w)) = H(w)^{-1},$$
the first equality because $H \equiv H^*$, the second equality by Theorem 3.3.4.

The primal-dual methods we present in sections 3.7 and 3.8 require that when given $x, s \in K^\circ$, one can compute the scaling point w as well as $g(w)$ and $H(w)$. In this regard, we note that if K is the nonnegative orthant and $e = (1, \ldots, 1)$, then w is the vector whose jth component is $\sqrt{x_j/s_j}$. Similarly, if X and S are pd matrices, and if E is the identity matrix, then
$$W = X^{1/2}(X^{1/2} S X^{1/2})^{-1/2} X^{1/2}.$$

3.5. Self-Scaled (or Symmetric) Cones

To get a sense of the role the scaling point plays in the theory of primal-dual methods, recall our discussion in §3.1 regarding the dependence of the primal and dual instances on the inner product. If in the initial inner product $\langle\ ,\ \rangle = \langle\ ,\ \rangle_e$ the primal instance is

$$\begin{aligned} \min\quad & \langle c, x \rangle \\ \text{s.t.}\quad & Ax = b, \\ & x \in K \end{aligned}$$

and the dual instance is

$$\begin{aligned} \max\quad & \langle b, y \rangle \\ \text{s.t.}\quad & A^*y + s = c, \\ & s \in K \quad (= K^*), \end{aligned}$$

then in terms of the local inner product $\langle\ ,\ \rangle_w$ the primal instance is

$$\begin{aligned} \min\quad & \langle c_w, x \rangle_w \\ \text{s.t.}\quad & Ax = b, \\ & x \in K \end{aligned}$$

and the dual instance is

$$\begin{aligned} \max\quad & \langle b, y \rangle \\ \text{s.t.}\quad & A_w^* y + s_w = c_w, \\ & s_w \in K, \end{aligned}$$

where

$$c_w := H(w)^{-1} c \quad \text{and} \quad A_w^* := H(w)^{-1} A^*.$$

Consequently, assuming w is the scaling point for a pair x, s, by rewriting the instances in terms of $\langle\ ,\ \rangle_w$, the pair is transformed to the pair $x, s_w := H(w)^{-1} s$ which satisfies $x = s_w$, i.e., each point in the pair is transformed to the point of intersection for the rewritten primal and dual feasible regions. The primal-dual methods we consider are invariant under the transformation in the sense that the iterates they generate for the rewritten instances are the transformations of the iterates generated for the original instances. For the rewritten instances, the primal and dual search directions—the directions in which one moves from the first and second points in the current pair x, s_w to the first and second points in the next pair—are respectively obtained by projecting the gradient (w.r.t. $\langle\ ,\ \rangle_w$) of a certain functional onto the nullspace of A and onto the range space of A_w^*. As these spaces are orthogonal complements (w.r.t. $\langle\ ,\ \rangle_w$), the sum of the search directions is the gradient itself. As we shall see, this makes for a particularly clean and symmetric analysis of the methods. However, before we can develop primal-dual methods, we need to know more about self-scaled cones.

Toward proving Theorem 3.5.3 we have the following proposition. *We suggest that the reader skip arguments marked "Proof" on the first pass through this subsection.*

Proposition 3.5.4. *Assume f is a barrier functional with $D_f = K°$, the interior of a (not necessarily self-scaled) cone. Let K^* denote the dual cone of K w.r.t. an arbitrary inner product $\langle\ ,\ \rangle$. If $x \in K°$ and $s \in (K°)^*$, then there exists $w \in K°$ such that $H(w)x = s$.*

Proof. Consider the functional

$$\psi(w) := \langle x, -g(w) \rangle + \langle s, w \rangle.$$

with domain K°. To prove the proposition, it suffices to show ψ has a minimizer, for at a minimizer the gradient is 0, that is, $-H(w)x + s = 0$. To prove ψ has a minimizer, it suffices to show that if $\{w_i\} \subset K^\circ$ is a sequence satisfying either $\|w_i\| \to \infty$ or $w_i \to \bar{w} \in \partial K$, then $\psi(w_i) \to \infty$.

Assume $\|w_i\| \to \infty$. Since $s \in (K^*)^\circ$ and $\{w_i\} \subset K$ it follows that $\langle s, w_i \rangle \to \infty$. Observing that $\langle x, -g(w_i) \rangle \geq 0$ because $-g(w_i) \in K^*$, we conclude $\psi(w_i) \to \infty$.

Now assume $w_i \to \bar{w} \in \partial K$. Theorem 2.2.9 shows $\|g(w_i)\| \to \infty$. Since $-g(w_i) \in K^*$ and $x \in K^\circ$ we thus have $\langle x, -g(w_i) \rangle \to \infty$. Observing that $\langle s, w_i \rangle \geq 0$ since $s \in K^*$ and $w_i \in K$, we conclude $\psi(w_i) \to \infty$. □

Applying Proposition 3.5.4 to a self-scaled cone K with a local inner product $\langle\,,\,\rangle = \langle\,,\,\rangle_e$, we have for each ordered pair $x, s \in K^\circ$ the existence of a scaling point w for the pair. In light of the discussion just prior to the statement of Theorem 3.5.3, to complete the proof of that theorem it remains only to prove the uniqueness of the scaling point. The uniqueness is crucial for the subsequent theory. To prove uniqueness, we rely on the following theorem which is important in itself. We freely invoke its identities throughout the remainder of the chapter; thus, the reader should keep them in mind.

Theorem 3.5.5. *Assume K is self-scaled. If $x, w \in K$, then*

$$f(H(w)x) = f(x) + 2(f(w) - f(e)), \tag{3.34}$$

$$g(H(w)x) = H(w)^{-1}g(x), \tag{3.35}$$

and

$$H(H(w)x) = H(w)^{-1}H(x)H(w)^{-1}. \tag{3.36}$$

Proof. Letting $C := C_e$, where C_e is as in the definition of an intrinsically self-conjugate functional, we have

$$\begin{aligned} f(H(w)x) &= f^*(H(w)x) - C \\ &= -\inf_{v \in K}(\langle v, H(w)x\rangle + f(v)) - C \\ &= -\inf_{v \in K}(\langle v, x\rangle_w + f(v)) - C \\ &= f_w^*(x) - C \\ &= f(x) + C_w - C. \end{aligned}$$

Using the values for C and C_w given by Proposition 3.5.1 establishes (3.34).

To prove (3.35), one can simply differentiate both sides of (3.34) w.r.t. x, making use of the chain rule. Alternatively, we can observe that since $-g$ is an involution (Theorem 3.5.2),

$$-g(-g(H(w)x)) = H(w)x.$$

Thus,

$$-H(w)^{-1}g(-g(H(w)x) = x,$$

that is,

$$-g_w(-g(H(w)x)) = x.$$

3.5. Self-Scaled (or Symmetric) Cones

Consequently, since $-g_w$ is an involution,
$$-g(H(w)x) = -g_w(x) = -H(w)^{-1}g(x),$$
establishing (3.35).

To establish (3.36), one can differentiate both sides of (3.35) w.r.t. x using the chain rule. Alternatively, relying on (3.35) and $-g$ being an involution, observe
$$\begin{aligned} H(H(w)x) &= H(-g(H(w)x))^{-1} \\ &= H(-H(w)^{-1}g(x))^{-1} \\ &= H(-g_w(x))^{-1} \\ &= H(-g_w(x))^{-1}H(w)H(w)^{-1} \\ &= H_w(-g_w(x))^{-1}H(w)^{-1} \\ &= H_w(x)H(w)^{-1} \\ &= H(w)^{-1}H(x)H(w)^{-1}. \end{aligned}$$

The proof is complete. $\square$

Proof of Theorem 3.5.3. As observed just prior to the statement of Theorem 3.5.5, it remains only to prove uniqueness of the scaling point for the ordered pair x, s.

Assume $H(w_1)x = s$ and $H(w_2)x = s$. By (3.36),
$$H(s)H(x) = (H(w_1)^{-1}H(x))^2 = (H(w_2)^{-1}H(x))^2$$
and so
$$(H(x)^{-1}H(w_1))^2 = (H(x)^{-1}H(w_2))^2,$$
that is,
$$H_x(w_1)^2 = H_x(w_2)^2.$$
Since the square root of a pd linear operator is unique, it follows that $H_x(w_1) = H_x(w_2)$ and hence
$$H_{w_1}(w_2) = H_x(w_1)^{-1}H_x(w_2) = I.$$
Thus,
$$g_{w_2} \equiv H_{w_1}(w_2)^{-1}g_{w_1} \equiv g_{w_1}.$$
In particular, relying on the logarithmic homogeneity of f,
$$w_2 = -g_{w_2}(w_2) = -g_{w_1}(w_2).$$
Of course logarithmic homogeneity also gives
$$w_1 = -g_{w_1}(w_1).$$

Consider the functional
$$w \mapsto \tfrac{1}{2}\|w\|_{w_1}^2 + f(w).$$
With respect to $\langle \ , \ \rangle_{w_1}$, the gradient at w is $w + g_{w_1}(w)$. Since $w_1 = -g_{w_1}(w_1)$ and $w_2 = -g_{w_1}(w_2)$, the gradient is 0 at both w_1 and w_2. As the functional is strictly convex and hence has at most one minimizer, $w_1 = w_2$. $\square$

3.5.4 Gradients and Norms

We continue to fix an arbitrary point $e \in K^\circ$ and let $\langle \, , \, \rangle = \langle \, , \, \rangle_e$.

The next theorem is of paramount importance. It concerns the error of the linear map $x \mapsto x - e$ in approximating the gradient map $x \mapsto g(x) - g(e)$. In terms of the norm $\| \, \| = \| \, \|_e$, we know from Proposition 2.2.10 that for self-concordant functionals, the error is small when $x \approx e$. The next theorem gives insight into the error for all x, not just for x near e. Its proof depends heavily on the uniqueness given by Theorem 3.5.3.

Recall that logarithmic homogeneity implies $g(e) = -e$.

Theorem 3.5.6. *Assume K is self-scaled. If $x \in K^\circ$, then*
$$(x - e) - (g(x) - g(e)) \in K,$$
that is,
$$-g(x) \in e - (x - e) + K.$$

Proof. Since $K^* = K$, it suffices to show for each $v \in K^\circ$ that
$$\langle (x - e) - (g(x) - g(e)), v \rangle \geq 0,$$
which is the same as
$$\psi(x) \geq \psi(e), \quad \text{where} \quad \psi(x) := \langle x - g(x), v \rangle.$$
Hence, it suffices to show e minimizes the functional ψ.

Toward proving e is indeed the minimizer, we first note ψ has a minimizer, as is immediate from two facts; if $\{x_i\} \subset K^\circ$ satisfies $\|x_i\| \to \infty$, then $\psi(x_i) \to \infty$, and if the sequence satisfies $x_i \to \bar{x} \in \partial K$, then $\psi(x_i) \to \infty$. The first fact follows from the relations $v \in K^\circ = (K^*)^\circ$, $-g(x_i) \in K^* = K$, whereas the second fact follows from the same relations along with $\|g(x_i)\| \to \infty$ if $x_i \to \bar{x} \in \partial K$, a consequence of Theorem 2.2.9. Hence, ψ has a minimizer x'.

At the mimimizer x', the gradient is 0, that is,
$$v - H(x')v = 0.$$
Since $H(e)v = Iv = v$, the uniqueness given by Theorem 3.5.3 thus implies $x' = e$. □

By definition, the local norms induced by a self-concordant functional f change gradually; if $\|x - e\| < 1$, then for all v,
$$1 - \|x - e\| \leq \frac{\|v\|_x}{\|v\|} \leq \frac{1}{1 - \|x - e\|}. \tag{3.37}$$

Stronger bounds hold if f is intrinsically self-conjugate, as we now describe.

Let $\mathcal{B}$ denote the largest set which is centrally symmetric about 0 and which satisfies $e + \mathcal{B} \subseteq K^\circ$, that is,
$$\mathcal{B} := \{v : e \pm v \in K^\circ\}.$$
Define a norm $| \, |$ on $\mathbb{R}^n$ by
$$|v| := \inf\{t \geq 0 : \tfrac{1}{t} v \in \mathcal{B}\}$$

3.5. Self-Scaled (or Symmetric) Cones

(the Minkowski gauge function for $\mathcal{B}$). For example, if $K = \mathbb{R}^n_+$ and $e = (1, \ldots, 1)$, then $|\ |$ is the ℓ_∞-norm, whereas $\|\ \|$ is the ℓ_2-norm.

Since $B(e, 1) := \{x : \|x - e\| < 1\}$ is a set which is centrally symmetric about e and satisfies $B(e, 1) \subseteq K^\circ$, we have $B(e, 1) \subseteq e + \mathcal{B}$. Stated differently,

$$|v| \leq \|v\| \quad \text{for all } v.$$

Theorem 3.5.7. *Assume K is self-scaled. If x satisfies $|x - e| < 1$, then for all v,*

$$\frac{1}{1 + |x - e|} \leq \frac{\|v\|_x}{\|v\|} \leq \frac{1}{1 - |x - e|} \tag{3.38}$$

and

$$1 - |x - e| \leq \frac{\|v\|_{-g(x)}}{\|v\|} \leq 1 + |x - e|. \tag{3.39}$$

The bounds (3.38) imply the Hessian of f to vary gradually. For example, just as we established the bounds in Theorem 2.2.1 using the bounds in the definition of self-concordance, from the bounds (3.38) we obtain that for all x satisfying $|x - e| < 1$,

$$\|I - H(x)\| \leq \frac{1}{(1 - |x - e|)^2} - 1. \tag{3.40}$$

This has implications for Newton's method when applied to functionals obtained by adding a linear term to f, like those for the barrier method (§2.4.1) and many other ipm's. To see why, assume z minimizes such a functional. Repeating the proof of Theorem 2.2.3, now making use of (3.40), we find that whenever $|z - e| < 1$, then one step of Newton's method beginning at e gives a point e_+ satisfying

$$\|z - e_+\| \leq \|z - e\| \frac{|z - e|}{1 - |z - e|}. \tag{3.41}$$

Compare this with the bound of Theorem 2.2.3:

$$\|z - e_+\| \leq \|z - e\| \frac{\|z - e\|}{1 - \|z - e\|}.$$

Unfortunately, although the stronger bound (3.41) indicates that ipm's like the barrier method might perform better when the underlying barrier functional is intrinsically self-conjugate, the stronger bound does not appear to be sufficient to prove complexity bounds which are better than what we have already proven, i.e., (2.18). Still, the bound is considered to be important conceptually and so we have discussed it.

The proof of Theorem 3.5.7 depends on the following theorem, which is of interest in itself and which plays a major role in the proof of an important theorem in §3.5.5 (a theorem which is central in our analysis of primal-dual methods).

Theorem 3.5.8. *Assume K is self-scaled. If x satisfies $x - e \in K$, then*

$$\|v\|_x \leq \|v\| \quad \text{for all } v.$$

Proof. We first establish the less restrictive inequality

$$\|v\|_x \leq (4\vartheta_f + 1)\|v\|. \tag{3.42}$$

For observe that $x - e \in K$ and $B(e, 1) \subseteq K$ (by self-concordance) together imply $B(x, 1) \subseteq K$ (keeping in mind that $B(x, 1) := B_e(x, 1)$). Hence, by symmetry and the fact that for each vector v, either $\langle g(x), v \rangle \geq 0$ or $\langle g(x), -v \rangle \geq 0$, Theorem 2.3.4 implies

$$B(x, 1) \subseteq B_x(x, 4\vartheta_f + 1),$$

giving (3.42).

Since

$$\|H(x)\|^{1/2} = \sup_v \frac{\|v\|_x}{\|v\|},$$

the inequality of the theorem is equivalent to asserting $\|H(x)\| \leq 1$.

Let $x_0 := x$ and $x_1 := H(x_0)^{-1}e$. Since logarithmic homogeneity implies $g(e) = -e$, we have $x_1 = -g_{x_0}(e)$. Hence, Theorem 3.5.6 applied with x_0 in place of e and e in place of x implies

$$e + x_1 - 2x_0 \in K.$$

Since $x_0 - e \in K$ it follows that $x_1 - x_0 \in K$ and hence $x_1 - e \in K$. Moreover, by (3.36) and $H(x_0)^{-1} = H(-g(x_0))$,

$$H(x_1) = H(H(x_0)^{-1}e) = H(x_0)H(e)H(x_0) = H(x_0)^2.$$

Proceeding inductively we find $x_k := H(x_{k-1})^{-1}e$ satisfies $x_k - e \in K$ and

$$H(x_k) = H(x)^{2^k}.$$

Thus, $\|H(x_k)\| = \|H(x)\|^{2^k}$. However, applying (3.42) with x_k in place of x, we have $\|H(x_k)\| \leq (4\vartheta_f + 1)^2$. Consequently, $\|H(x)\| \leq 1$. □

Proof of Theorem 3.5.7. Let

$$x_1 := \frac{1}{1 + |x - e|}x \quad \text{and} \quad x_2 := e - \frac{1}{|x - e|}(x - e).$$

Since $|x_2 - e| = 1$ we have $x_2 \in K$. Observe that

$$e = x_1 + \frac{|x - e|}{1 + |x - e|}x_2.$$

Hence, $e - x_1 \in K$. Consequently, applying Theorem 3.5.8 with x_1 in place of e and e in place of x,

$$\|v\| \leq \|v\|_{x_1} \quad \text{for all } v.$$

Since logarithmic homogeneity of f makes the Hessian homogeneous of degree -2, we thus conclude by definition of x_1 that

$$\|v\| \leq (1 + |x - e|)\|v\|_x \quad \text{for all } v,$$

3.5. Self-Scaled (or Symmetric) Cones

giving the leftmost inequality in (3.38).

Similarly, to establish the rightmost inequality in (3.38), define

$$x_1' := (1 - |x - e|)e \quad \text{and} \quad x_2' := e + \frac{1}{|x - e|}(x - e) \in K.$$

Note $x_2' \in K$ and $x = x_1' + |x - e|x_2'$. Hence, $x - x_1' \in K$. Theorem 3.5.8 thus implies for all v,

$$\|v\|_x \leq \|v\|_{x_1'}.$$

Logarithmic homogeneity gives

$$\|v\|_x \leq \frac{1}{1 - |x - e|}\|v\| \quad \text{for all } v,$$

establishing the desired inequality.

To prove (3.39) one uses (3.38) and the relation $H(-g(x)) = H(x)^{-1}$ (so the eigenvalues of $H(-g(x))$ are the reciprocals of the eigenvalues of $H(x)$). □

Recall Theorem 2.3.4 which asserts that if f is a barrier functional and if $x, y \in D_f$ satisfy $\langle g(x), y - x \rangle \geq 0$, then $y \in B_x(x, 4\vartheta_f + 1)$. Among other things, this implies that the set $B_x(x, 1)$ is, to within a factor $4\vartheta_f + 1$, the largest among all ellipsoids centered at x which are contained in D_f. A stronger statement holds when f is intrinsically self-conjugate.

Theorem 3.5.9. *Assume K is self-scaled. If $x \in K^\circ$ satisfies $\langle g(e), x - e \rangle \geq 0$, then $x \in B(e, \vartheta_f)$.*

Proof. Assume $x \in K^\circ$ satisfies $\langle g(e), x - e \rangle \geq 0$. Since $g(e) = -e$, we are thus assuming $\langle x, e \rangle \leq \|e\|^2$, an upper bound on $\langle x, e \rangle$. A strict lower bound of 0 is immediate from $e, x \in K^\circ = (K^*)^\circ$. In all,

$$0 < \langle x, e \rangle \leq \|e\|^2.$$

Thus, the point

$$\bar{x} := \frac{\|e\|^2}{\langle x, e \rangle} x$$

is a scalar multiple of x where the scalar is no less than 1. Stated differently, x is a convex combination of 0 and $\bar{x}$. Hence,

$$\begin{aligned}\|x - e\| &\leq \max\{\|0 - e\|, \|\bar{x} - e\|\} \\ &= \max\{\sqrt{\vartheta_f}, \|\bar{x} - e\|\},\end{aligned} \quad (3.43)$$

the equality by logarithmic homogeneity (Theorem 2.3.9).

Let

$$v := \frac{1}{\|\bar{x} - e\|}(\bar{x} - e).$$

Since $\|v\| = 1$ we have $e - v \in K$. Since $K^* = K$, we thus see $\langle \bar{x}, e - v \rangle \geq 0$, that is,

$$\langle e + \|\bar{x} - e\|v, e - v \rangle \geq 0.$$

Hence, using $\|e\|^2 = \vartheta_f$, $\langle v, e \rangle = 0$, and $\|v\| = 1$, we have $\|\bar{x} - e\| \le \vartheta_f$. Consequently, by (3.43) we find $\|x - e\| \le \vartheta_f$. By openness, the inequality is seen to be strict, that is, $x \in B(e, \vartheta_f)$. $\square$

Theorem 3.5.10. *Assume K is self-scaled and $x \in K^\circ$. For each $t \in \mathbb{R}$ there exists unique $x_t \in K^\circ$ such that*
$$H(x_t) = H(x)^t.$$

Proof. That there is at most one point x_t with the desired property is immediate from the uniqueness in Theorem 3.5.3.

Define
$$T := \{t \in \mathbb{R} : \exists x_t \in K^\circ \text{ s.t. } H(x_t) = H(x)^t\}.$$
It suffices to prove that T is a closed and dense subset of $\mathbb{R}$.

For proving that T is closed, assume t to be in the closure of T and let $\{t_i\} \subset T$ be a sequence satisfying $t_i \to t$. We claim the sequence $\{x_{t_i}\}$ has a subsequence converging to a point $x' \in K^\circ$. Thus, by continuity of the Hessian, $H(x') = H(x)^t$. Hence $t \in T$ and $x_t := x'$, establishing closedness.

To show $\{x_{t_i}\}$ has a subsequence converging to a point $x' \in K^\circ$, it suffices to show the sequence is bounded and does not have any subsequence converging to a point in the boundary of K. If the sequence (similarly, a subsequence) satisfied $\|x_{t_i}\| \to \infty$, then using $H(x_{t_i}) \to H(x)^t$ we could conclude
$$\|x_{t_i}\|_{x_{t_i}}^2 = \langle x_{t_i}, H(x)^{t_i} x_{t_i} \rangle \to \infty,$$
contradicting $\|x_{t_i}\|_{x_{t_i}}^2 = \vartheta_f$ (Theorem 2.3.9). If the sequence (similarly, a subsequence) converged to a point in the boundary of K, then from the containments $B_{x_{t_i}}(x_{t_i}, 1) \subseteq K^\circ$ we could conclude $\|H(x_{t_i})\| \to \infty$, contradicting $H(x_{t_i}) \to H(x)^t$. Hence, $\{x_{t_i}\}$ does indeed have a subsequence converging to a point $x' \in K^\circ$, completing the proof that T is closed.

To prove that T is dense in $\mathbb{R}$, we show T contains all quantities that can be obtained by adding finitely many numbers each of which is an integral power of 2 and contains the negatives of those quantities. Noting $0, 1 \in T$ since $H(e) = I = H(x)^0$ and $H(x) = H(x)^1$, for this it suffices to establish the following: (i) if $t \in T$, then $-t, 2t \in T$ and (ii) if $t_1, t_2 \in T$, then $\frac{1}{2}(t_1 + t_2) \in T$.

Assume $t \in T$. Since $H(-g(x_t)) = H(x_t)^{-1} = H(x)^{-t}$ we have $-t \in T$. Defining $x_{2t} := H(-g(x_t))e = H(x_t)^{-1}e$, (3.36) implies $H(x_{2t}) = H(x_t)^2 = H(x)^{2t}$. Hence, $2t \in T$. We have proved (i).

In proving (ii), we rely on the fact that for each inner product $\langle\,,\,\rangle$ and for each pd linear operator H, if $t_1, t_2 \in \mathbb{R}$, then H^{t_1} is pd w.r.t. the inner product defined by $u, v \mapsto \langle u, H^{t_2}v \rangle$. We leave the proof of this fact to the reader.

Assume $t_1, t_2 \in T$. We know $-t_2 \in T$. Let w be the scaling point for the pair x_{t_1}, x_{-t_2}, that is, $H(w)x_{t_1} = x_{-t_2}$. Then (3.36) implies
$$(H(w)^{-1}H(x_{t_1}))^2 = H(x_{-t_2})H(x_{t_1}),$$
that is,
$$H_{x_{t_1}}(w)^{-2} = H(x)^{t_1 - t_2}. \tag{3.44}$$

3.5. Self-Scaled (or Symmetric) Cones

By the fact mentioned above, the operators $H(x)^{t_1-t_2}$ and $H(x)^{\frac{1}{2}(t_1-t_2)}$ are pd w.r.t. $\langle\,,\,\rangle_{x_{t_1}}$. Since the square root of a pd operator is unique, (3.44) thus implies

$$H_{x_{t_1}}(w) = H(x)^{\frac{1}{2}(t_2-t_1)}.$$

Consequently,

$$H(w) = H(x_{t_1})H_{x_{t_1}}(w) = H(x)^{t_1}H(x)^{\frac{1}{2}(t_2-t_1)} = H(x)^{\frac{1}{2}(t_1+t_2)},$$

so that $\frac{1}{2}(t_1+t_2) \in T$. □

3.5.5 A Useful Theorem

We continue to fix an arbitrary point $e \in K^\circ$ and let $\langle\,,\,\rangle = \langle\,,\,\rangle_e$.

We close this section by proving a theorem which plays a central role in the analysis of the primal-dual methods we consider.

Theorem 3.5.11. *Assume K is a self-scaled cone. If $x \in K^\circ$, then*

$$\frac{\|x+g(x)\|}{\max\{\|H(x)^{1/2}\|, \|H(x)^{-1/2}\|\}} \geq \min\{\tfrac{1}{5}, \tfrac{4}{5}\|x-e\|\}.$$

The following theorem and proposition are used to prove Theorem 3.5.11.

Theorem 3.5.12. *Assume K is a self-scaled cone. If $x \in K^\circ$ and*

$$\lambda := \max\{\|H(x)\|, \|H(x)^{-1}\|\},$$

then

$$\|x+g(x)\| \geq \frac{\lambda-1}{\sqrt{\lambda}}.$$

Proof. We may assume $\lambda > 1$.

Consider the case that $\lambda = \|H(x)\|$, i.e., λ is the largest eigenvalue of $H(x)$. For $0 \leq \epsilon < \lambda$ let

$$z_\epsilon := \tfrac{-1}{\lambda-\epsilon}g(x).$$

Since

$$\begin{aligned}H_{z_\epsilon}(x) &= H(z_\epsilon)^{-1}H(x)\\ &= \tfrac{1}{(\lambda-\epsilon)^2}H(-g(x))^{-1}H(x)\\ &= \tfrac{1}{(\lambda-\epsilon)^2}H(x)^2,\end{aligned}$$

the largest eigenvalue of $H_{z_\epsilon}(x)$ is $(\lambda/(\lambda-\epsilon))^2$; in particular, if $0 < \epsilon < \lambda$, the largest eigenvalue is greater than 1. Hence, if $0 < \epsilon < \lambda$, Theorem 3.5.8 (applied with z_ϵ in place of e) shows

$$x - z_\epsilon \notin K,$$

that is,
$$x + \tfrac{1}{\lambda-\epsilon} g(x) \notin K.$$

Equivalently (assuming $\epsilon \neq \lambda - 1$),
$$\tfrac{\lambda-1-\epsilon}{\lambda-\epsilon}\left(x + \tfrac{1}{\lambda-1-\epsilon}(x + g(x))\right) \notin K.$$

Since K is a cone, if $0 < \epsilon < \lambda - 1$ we thus have
$$x + \tfrac{1}{\lambda-1-\epsilon}(x + g(x)) \notin K.$$

Consequently, since ϵ can be arbitrarily near 0 and since $B_x(x, 1) \subseteq K^\circ$ (by self-concordance),
$$\|x + g(x)\|_x \geq \lambda - 1,$$

implying the desired inequality
$$\|x + g(x)\| \geq \frac{\lambda - 1}{\sqrt{\lambda}}.$$

In the case that $\lambda = \|H(x)^{-1}\| \, (= \|H(-g(x))\|)$, one simply interchanges the roles of x and $-g(x)$. $\square$

Proposition 3.5.13. *Assume f is a logarithmically homogeneous barrier functional with domain K°. If $x \in K^\circ$, then*
$$\|x + g(x)\| \geq \min\left\{\tfrac{1}{4}, \|x - e\|\right\}.$$

Proof. Consider the self-concordant functional
$$f'(x) := \tfrac{1}{2}\|x\|^2 + f(x),$$

where $\|\;\|$ is the local norm at e given by f. Note the following relation between the local inner product at e given by f' and the one given by f: $\langle\;,\;\rangle' = 2\langle\;,\;\rangle$. Letting g' denote the gradient of f' w.r.t. $\langle\;,\;\rangle'$, it follows that
$$g'(x) = \tfrac{1}{2}(x + g(x)).$$

Since f is logarithmically homogeneous, $g(e) = -e$ and hence $g'(e) = 0$. Applying Proposition 2.2.10 to f', we thus find that for each value $r \leq \tfrac{1}{4}$, the image of the closed ball $\bar{B}'(e, r + \tfrac{3r^2}{(1-r)^3})$ under the map $x \mapsto g'(x)$ contains the closed ball $\bar{B}'(0, r)$. In particular, if $r \leq \tfrac{1}{4\sqrt{2}}$ (then $\tfrac{3r}{(1-r)^3} < 1$), the image of $\bar{B}'(e, 2r)$ contains $\bar{B}'(0, r)$. Since the map $x \mapsto g'(x)$ is injective (as are the gradient maps for all strictly convex functionals), it follows that for all $x \in K^\circ$,
$$\|g'(x)\|' \geq \min\left\{\tfrac{1}{4\sqrt{2}}, \tfrac{1}{2}\|x - e\|'\right\},$$

i.e., it follows that
$$\tfrac{1}{\sqrt{2}}\|x + g(x)\| \geq \min\left\{\tfrac{1}{4\sqrt{2}}, \tfrac{1}{\sqrt{2}}\|x - e\|\right\},$$

3.6. The Nesterov–Todd Directions

proving the proposition. □

Proof of Theorem 3.5.11. Theorem 3.5.12 and Proposition 3.5.13 give

$$\frac{\|x + g(x)\|}{\max\{\|H(x)^{1/2}\|, \|H(x)^{-1/2}\|\}} \geq \max\left\{\frac{\lambda - 1}{\lambda}, \frac{\beta}{\sqrt{\lambda}}\right\}, \tag{3.45}$$

where

$$\beta := \min\left\{\tfrac{1}{4}, \|x - e\|\right\}.$$

Note that $\delta \mapsto (\delta - 1)/\delta$ is an increasing function of δ when $\delta > 0$, whereas for fixed $\gamma > 0$, the function $\delta \mapsto \gamma/\sqrt{\delta}$ is decreasing. Consequently, the function

$$\delta \mapsto \max\left\{\frac{\delta - 1}{\delta}, \frac{\gamma}{\sqrt{\delta}}\right\}$$

is minimized when δ satisfies

$$\frac{\delta - 1}{\delta} = \frac{\gamma}{\sqrt{\delta}},$$

that is, when

$$\delta = \left(\frac{\gamma + \sqrt{\gamma^2 + 4}}{2}\right)^2 \quad \text{and the function value is} \quad \frac{2\gamma}{\gamma + \sqrt{\gamma^2 + 4}}.$$

Thus, by (3.45),

$$\frac{\|x + g(x)\|}{\max\{\|H(x)^{1/2}\|, \|H(x)^{-1/2}\|\}} \geq \frac{2\beta}{\beta + \sqrt{\beta^2 + 4}}$$

$$\geq \frac{\beta}{1 + \beta}$$

$$\geq \frac{\min\left\{\tfrac{1}{4}, \|x - e\|\right\}}{1 + \tfrac{1}{4}},$$

completing the proof. □

3.6 The Nesterov–Todd Directions

In moving from a primal and dual feasible pair $x, (y, s)$ to the next feasible pair $x_+, (y_+, s_+)$, primal-dual ipm's rely on both x and (y, s) in determining the primal direction $x_+ - x$ and in determining the dual direction $(y_+ - y, s_+ - s)$. Here we study what are perhaps the most prevalent directions appearing in the design of primal-dual ipm's.

We continue to assume K is a self-scaled cone. We fix an arbitrary point $e \in K^\circ$ and let $\langle\ ,\ \rangle = \langle\ ,\ \rangle_e$; that is, we continue to suppress the subscript e.

The primal instance is

$$\begin{aligned}
\min\ & \langle c, x \rangle \\
\text{s.t.}\ & Ax = b, \\
& x \in K
\end{aligned}$$

and the dual instance is

$$\begin{aligned} \max \quad & \langle b, y \rangle \\ \text{s.t.} \quad & A^*y + s = c, \\ & s \in K. \end{aligned}$$

More geometrically, we know from §3.1 that fixing $\hat{x}$ satisfying $A\hat{x} = b$ and $(\hat{y}, \hat{s})$ satisfying $A^*\hat{y} + \hat{s} = c$, and letting L denote the nullspace of A, the primal and dual instances can be expressed as

$$\begin{aligned} \min \quad & \langle \hat{s}, x \rangle \\ \text{s.t.} \quad & x \in L + \hat{x}, \\ & x \in K \end{aligned}$$

and

$$\begin{aligned} \min \quad & \langle \hat{x}, s \rangle \\ \text{s.t.} \quad & s \in L^\perp + \hat{s}, \\ & s \in K. \end{aligned}$$

The primal central path is the set

$$\{x_\eta : \eta > 0\},$$

where x_η solves

$$\begin{aligned} \min_x \quad & \eta \langle c, x \rangle + f(x) \\ \text{s.t.} \quad & Ax = b, \end{aligned} \tag{3.46}$$

that is, solves

$$\begin{aligned} \min_x \quad & \eta \langle \hat{s}, x \rangle + f(x) \\ \text{s.t.} \quad & x \in L + \hat{x}. \end{aligned}$$

The dual central path is the set

$$\{(y_\eta, s_\eta) : \eta > 0\},$$

where (y_η, s_η) solves

$$\begin{aligned} \max_{y,s} \quad & \eta \langle b, y \rangle - f(s) \\ \text{s.t.} \quad & A^*y + s = c. \end{aligned}$$

Thus, s_η solves

$$\begin{aligned} \min_s \quad & \eta \langle \hat{x}, s \rangle + f(s) \\ \text{s.t.} \quad & s \in L^\perp + \hat{s}. \end{aligned}$$

To highlight symmetry between the primal and dual instances, we often hide y_η, referring to $\{s_\eta : \eta > 0\}$ as the dual central path, keeping in mind that y_η is uniquely determined by s_η through the equations $A^*y_\eta + s_\eta = c$ (assuming A is surjective and hence A^* is injective).

Assume x approximates x_η and (y, s) approximates (y_η, s_η). One strategy for obtaining a better approximation to x_η is to move from x in the steepest descent direction for the linearly constrained optimization problem (3.46), i.e., in the direction opposite the projected gradient. This direction w.r.t. the inner product $\langle \ , \ \rangle_x$ is the direction of the Newton step, the step taken by the barrier method. If, instead, one relies on $\langle \ , \ \rangle_w$ where w is the scaling point for the ordered pair x, s, one arrives at the *primal Nesterov–Todd direction*:

$$d_x := -P_{L,w}(\eta c_w + g_w(x)),$$

3.6. The Nesterov–Todd Directions

where $c_w := H(w)^{-1}c$ and $P_{L,w}$ denotes orthogonal projection onto L, orthogonal w.r.t. $\langle\ ,\ \rangle_w$. Since $c_w = A_w^* y + H(w)^{-1}s$ and $P_{L,w}A_w^* \equiv 0$ (where $A_w^* = H(w)^{-1}A^*$ is the adjoint of A w.r.t. $\langle\ ,\ \rangle_w$), the direction can also be expressed as

$$d_x = -P_{L,w}(\eta H(w)^{-1}s + g_w(x))$$
$$= -P_{L,w}(\eta x + g_w(x)).$$

The *dual Nesterov–Todd direction* is

$$d_s := -P_{L^\perp, w^*}(\eta H(w^*)^{-1}x + g_{w^*}(s))$$
$$= -P_{L^\perp, w^*}(\eta s + g_{w^*}(s)),$$

where $w^* = -g(w)$ is the scaling point for the ordered pair s, x (and where $L^\perp$ is the orthogonal complement of L w.r.t. the *original* inner product $\langle\ ,\ \rangle$).

One gets a primal Nesterov–Todd direction and a dual Nesterov–Todd direction for each triple η, x, s, where x is primal feasible and s is dual feasible.

The Nesterov–Todd directions appeared in primal-dual ipm's for LP long before Nesterov and Todd developed them generally (cf. Megiddo [13], Kojima, Mizuno, and Yoshise [11], and Monteiro and Adler [14]). Indeed, the goal of Nesterov and Todd was to generalize earlier work for LP.

Theoretical and practical insight into the Nesterov–Todd directions is gained by considering d_x together with $H(w)^{-1}d_s$ (or, symmetrically, $H(w^*)^{-1}d_x$ together with d_s).

Note that $H(w)^{-1}d_s \in L^{\perp_w}$, where $L^{\perp_w} = H(w)^{-1}L^\perp$ is the orthogonal complement of L w.r.t. $\langle\ ,\ \rangle_w$. We claim that

$$H(w)^{-1}d_s = -P_{L^{\perp_w}, w}(\eta x + g_w(x)), \tag{3.47}$$

and hence

$$d_x + H(w)^{-1}d_s = -(\eta x + g_w(x))$$

gives an $\langle\ ,\ \rangle_w$-orthogonal decomposition of $-(\eta x + g_w(x))$.

To establish (3.47) it suffices to prove the operator identity

$$P_{L^\perp, w^*} \equiv H(w) P_{L^{\perp_w}, w} H(w)^{-1}, \tag{3.48}$$

for then

$$H(w)^{-1}d_s = -H(w)^{-1}P_{L^\perp, w^*}(\eta s + g_{w^*}(s))$$
$$= -P_{L^{\perp_w}, w}H(w)^{-1}(\eta s + g_{w^*}(s))$$
$$= -P_{L^{\perp_w}, w}H(w)^{-1}(\eta s + H(w^*)^{-1}g(s))$$
$$= -P_{L^{\perp_w}, w}(\eta x + g(s))$$
$$= -P_{L^{\perp_w}, w}(\eta x + g(H(w)x))$$
$$= -P_{L^{\perp_w}, w}(\eta x + H(w)^{-1}g(x))$$
$$= -P_{L^{\perp_w}, w}(\eta x + g_w(x)).$$

To prove the operator identity (3.48) it suffices to show

$$\text{for all } u \in L^\perp, \quad H(w)P_{L^{\perp_w}, w}H(w)^{-1}u = u \tag{3.49}$$

and
$$\text{for all } v \in (L^\perp)^{\perp_{w^*}}, \quad H(w) P_{L^{\perp_w},w} H(w)^{-1} v = 0. \tag{3.50}$$

However, (3.49) is immediate from the identity $L^{\perp_w} = H(w)^{-1} L^\perp$, whereas (3.50) is immediate from

$$(L^{\perp_w})^{\perp_w} = L = (L^\perp)^\perp = H(w^*)(L^\perp)^{\perp_{w^*}} = H(w)^{-1}(L^\perp)^{\perp_{w^*}}.$$

Hence we have established (3.47), with the consequence that

$$d_x + H(w)^{-1} d_s = -(\eta x + g_w(x))$$

does indeed give an $\langle \, , \, \rangle_w$-orthogonal decomposition of $-(\eta x + g_w(x))$.

We reintroduce y, keeping in mind that y is uniquely determined by s and, likewise, the direction d_y is uniquely determined by d_s. We now see d_x, d_s, and d_y satisfy

$$\begin{bmatrix} 0 & I & A_w^* \\ I & I & 0 \\ A & 0 & 0 \end{bmatrix} \begin{bmatrix} d_x \\ H(w)^{-1} d_s \\ d_y \end{bmatrix} = - \begin{bmatrix} 0 \\ \eta x + g_w(x) \\ 0 \end{bmatrix};$$

that is, they satisfy

$$\begin{bmatrix} 0 & I & A^* \\ H(w) & I & 0 \\ A & 0 & 0 \end{bmatrix} \begin{bmatrix} d_x \\ d_s \\ d_y \end{bmatrix} = - \begin{bmatrix} 0 \\ \eta s + g(x) \\ 0 \end{bmatrix}.$$

In the literature, these equations are often taken as the starting point for discussions of the Nesterov–Todd directions, i.e., they are taken as defining the directions. The equations can be used to compute the directions in practice.

It should be mentioned that d_x and $H(w)^{-1} d_s$ are the Nesterov–Todd directions if one replaces $\langle \, , \, \rangle$ with $\langle \, , \, \rangle_w$, rewriting the instances in terms of this inner product. Moreover, one can show quite generally that the Nesterov–Todd directions are essentially invariant under changes in the inner product. For example, if one replaces $\langle \, , \, \rangle$ with $\langle \, , \, \rangle_z$ where $z \in K^\circ$, then d_x and $H(z)^{-1} d_s$ are the Nesterov–Todd directions for the rewritten instances

$$\begin{array}{ll} \min & \langle c_z, x \rangle_z \\ \text{s.t.} & Ax = b, \\ & x \in K, \end{array} \qquad \begin{array}{ll} \max & \langle b, y \rangle \\ \text{s.t.} & A_z^* y + s_z = c_z, \\ & s_z \in K. \end{array}$$

The simple geometry given by viewing the summation $d_x + H(w)^{-1} d_s$ as an $\langle \, , \, \rangle_w$-orthogonal decomposition of $-(\eta x + g_w(x))$ provides analyses of primal-dual ipm's which are more transparent than arguments phrased only in terms of the original inner product $\langle \, , \, \rangle$ and the original instances. However, other inner products and rewritings of the instances reveal the same simple geometry, for example, the approach which is symmetric to the one we employ. The symmetric approach relies on $\langle \, , \, \rangle_{w^*}$. In this approach, $L^\perp$ is fixed, L is replaced by $(L^\perp)^{\perp_{w^*}} = H(w^*)^{-1} L$, and the sum $H(w^*)^{-1} d_x + d_s$ gives an $\langle \, , \, \rangle_{w^*}$-orthogonal decomposition of $-(\eta s + g_{w^*}(s))$.

One can even use the original inner product $\langle \, , \, \rangle$ to reveal the same simple geometry *if* one transforms the instances appropriately. Specifically, transforming the primal feasible

3.6. The Nesterov–Todd Directions

region by $H(w)^{1/2} = H(w^*)^{-1/2}$ and the dual feasible region by $H(w^*)^{1/2} = H(w)^{-1/2}$, we see that the original primal and dual instances are, respectively, equivalent to the primal instance

$$\min_{x'} \quad \langle H(w^*)^{1/2}\hat{s}, x' \rangle$$
$$\text{s.t.} \quad x' \in H(w)^{1/2}(L + \hat{x}),$$
$$x' \in K \quad (= H(w)^{1/2} K)$$

and its dual instance

$$\min_{s'} \quad \langle H(w)^{1/2}\hat{x}, s' \rangle$$
$$\text{s.t.} \quad s' \in H(w^*)^{1/2}(L^\perp + \hat{s}),$$
$$s' \in K \quad (= H(w^*)^{1/2} K).$$

Both of the original points x, s are transformed to the point $z := H(w)^{1/2}x = H(w^*)^{1/2}s$, and $H(w)^{1/2}d_x + H(w^*)^{1/2}d_s$ gives an $\langle \ , \ \rangle$-orthogonal decomposition of $-(\eta z + g(z))$. This last approach—relying on the original inner product $\langle \ , \ \rangle$ but transforming the primal and dual instances—is perhaps more consistent with the literature than the approach we employ which uses the local inner product $\langle \ , \ \rangle_w$, but it has the drawback of messier notation and makes the appearance of w, w^* through $H(w)^{1/2}, H(w^*)^{1/2}$ unnecessarily mysterious. Consequently, we have chosen to rely on $\langle \ , \ \rangle_w$, but we emphasize that the various approaches are essentially equivalent.

In closing this section, we remark that in the analysis of algorithms using the Nesterov–Todd directions, it is convenient to focus on

$$\bar{w} := \tfrac{1}{\sqrt{\eta}} w \quad \text{and} \quad \bar{w}^* := \tfrac{1}{\sqrt{\eta}} w^*$$

rather than on w and w^*. (Take note that we slightly abuse notation in that $\bar{w}^* \neq -g(\bar{w})$ unless $\eta = 1$.) To give a sense as to why it is convenient, we show that a necessary and sufficient condition to have the simultaneous equalities $x = x_\eta$ and $s = s_\eta$ is that $\bar{w} = x$ (likewise, $\bar{w}^* = s$), for observe that $\bar{w}$ is the scaling point for the ordered pair $x, \eta s$ (whereas $\bar{w}^*$ is the scaling point for the pair $s, \eta x$). Hence,

$$\begin{aligned} g_{\bar{w}}(x) &= H(\bar{w})^{-1} g(x) \\ &= g(H(\bar{w})x) \\ &= g(\eta s) \\ &= \tfrac{1}{\eta} g(s). \end{aligned} \quad (3.51)$$

As we know from §3.4, a necessary and sufficient condition for the simultaneous equalities $x = x_\eta$ and $s = s_\eta$ is that $x = -\tfrac{1}{\eta} g(s)$. Consequently, since logarithmic homogeneity and the uniqueness of scaling points (Theorem 3.5.3) imply $g_{\bar{w}}(x) = -x$ iff $\bar{w} = x$, (3.51) shows we do indeed have the simultaneous equalities $x = x_\eta$ and $s = s_\eta$ iff $\bar{w} = x$.

In the remaining sections of this chapter it is useful to scale the Nesterov–Todd directions, letting

$$\bar{d}_x := \tfrac{1}{\eta} d_x \quad \text{and} \quad \bar{d}_s := \tfrac{1}{\eta} d_s.$$

Since $H(\bar{w}) = \eta H(w)$ and $H(\bar{w}^*) = \eta H(w^*)$, we have $P_{L,w} \equiv P_{L,\bar{w}}$, $P_{L^\perp, w^*} \equiv P_{L^\perp, \bar{w}^*}$,

$$\bar{d}_x = -P_{L,\bar{w}}(x + g_{\bar{w}}(x))$$

and
$$\bar{d}_s = -P_{L^\perp, \bar{w}^*}(s + g_{\bar{w}^*}(s)).$$

Moreover,
$$\bar{d}_x + H(w)^{-1}\bar{d}_s = -(x + g_{\bar{w}}(x))$$

is an $\langle\,,\,\rangle_{\bar{w}}$-orthogonal decomposition of $-(x + g_{\bar{w}}(x))$. (In the last equality, we wrote $H(w)^{-1}\bar{d}_s$ rather than $\eta H(\bar{w})^{-1}\bar{d}_s$ to ease notation.)

3.7 Primal-Dual Path-Following Methods

We assume K is a self-scaled cone and rely on the notation of §3.6. We continue to fix an arbitrary point $e \in K^\circ$ and let $\langle\,,\,\rangle := \langle\,,\,\rangle_e$.

3.7.1 Measures of Proximity

The analysis of primal-dual path-following ipm's relies on measuring proximity of current points x and (y, s) to the primal and dual central paths. In the literature, there are various expressions for measuring the proximity. Perhaps the most basic approach is simply to measure the distance from x to x_η by $\|x - x_\eta\|_{x_\eta}$ and, likewise, measure the distance from (y, s) to (y_η, s_η) by $\|s - s_\eta\|_{s_\eta}$. However, other expressions for measuring the proximity yield more elegant and insightful analyses of primal-dual methods.

Let
$$\bar{w} := \tfrac{1}{\sqrt{\eta}} w \quad \text{and} \quad \bar{w}^* := \tfrac{1}{\sqrt{\eta}} w^*,$$

where w is the scaling point for the ordered pair x, s and w^* ($= -g(w)$) is the scaling point for the pair s, x. We know from §3.6 that a necessary and sufficient condition for the simultaneous equalities $x = x_\eta$ and $s = s_\eta$ is that $\bar{w} = x$ (likewise, $\bar{w}^* = s$). Thus, one might consider measuring proximity of the pair x, s to the central path pair x_η, s_η by the quantity $\|x - \bar{w}\|_{\bar{w}}$. It is not difficult to show

$$\|x - \bar{w}\|_{\bar{w}} = \|s - \bar{w}^*\|_{\bar{w}^*},$$

so we have symmetry in this measure of proximity.

Although the quantity $\|x - \bar{w}\|_{\bar{w}}$ plays an important role in our analysis, it does not have the starring role. That role is played by the value $\|\eta s + g(x)\|_{-g(x)}$, the usefulness of which is indicated by the following theorem.

Recall that $\langle x, s\rangle = \langle c, x\rangle - \langle b, y\rangle$ for feasible points x and (y, s).

Theorem 3.7.1. *If $x, s \in K^\circ$, then*

$$\|\eta s + g(x)\|_{-g(x)} = \|\eta x + g(s)\|_{-g(s)} \tag{3.52}$$

and

$$|\langle x, s\rangle - \tfrac{1}{\eta}\vartheta_f| \le \tfrac{1}{\eta}\sqrt{\vartheta_f}\|\eta s + g(x)\|_{-g(x)}. \tag{3.53}$$

Moreover,

$$\|\eta s + g(x)\|_{-g(x)} \ge \min\left\{\tfrac{1}{5}, \tfrac{4}{5}\|x - \bar{w}\|_{\bar{w}}\right\}.$$

3.7. Primal-Dual Path-Following Methods

Proof. To prove (3.52), substitute $H(w)^{-1}s$ for x and $H(w)x$ for s in the expression $\|\eta s + g(x)\|_{-g(x)}$, then use the identities

$$g(H(w)^{-1}s) = H(w)g(s) \quad \text{and} \quad H(H(w)g(s)) = H(w)^{-1}H(g(s))H(w)^{-1}.$$

Next, we claim that $H(x)^{1/2}e = -g(x)$, for by Theorem 3.5.10 there exists $x_{1/2} \in K^\circ$ such that $H(x_{1/2}) = H(x)^{1/2}$ and hence by (3.36),

$$H(H(x)^{1/2}e) = H(x)^{-1} = H(-g(x)).$$

Thus, $H(x)^{1/2}e = -g(x)$ by the uniqueness given in Theorem 3.5.3. Likewise, $H(x)^{-1/2}e = x$. Consequently, relying on the fact that $\langle x, g(x) \rangle = -\vartheta_f$ by logarithmic homogeneity (Theorem 2.3.9), we have

$$\begin{aligned}
\left|\langle x, s\rangle - \tfrac{1}{\eta}\vartheta_f\right| &= \left|\left\langle x, s + \tfrac{1}{\eta}g(x)\right\rangle\right| \\
&= \left|\left\langle e, H(x)^{-1/2}\left(s + \tfrac{1}{\eta}g(x)\right)\right\rangle\right| \\
&\leq \|e\| \left\|H(x)^{-1/2}\left(s + \tfrac{1}{\eta}g(x)\right)\right\| \\
&= \tfrac{1}{\eta}\sqrt{\vartheta_f}\|H(x)^{-1/2}(\eta s + g(x))\| \\
&= \tfrac{1}{\eta}\sqrt{\vartheta_f}\|H(-g(x))^{1/2}(\eta s + g(x))\| \\
&= \tfrac{1}{\eta}\sqrt{\vartheta_f}\|\eta s + g(x)\|_{-g(x)}.
\end{aligned}$$

Toward proving the final inequality of the theorem, we remark that from the identities $H(\bar{w})^{-1}\eta s = x$, $g_{\bar{w}}(x) = H(\bar{w})^{-1}g(x)$, and

$$H(-H(\bar{w})^{-1}g(x)) = H(\bar{w})H(-g(x))H(\bar{w}),$$

one can readily show

$$\|\eta s + g(x)\|_{-g(x)} = \|x + g_{\bar{w}}(x)\|_{-g_{\bar{w}}(x)}.$$

Since for all v,

$$\|v\|_{-g_{\bar{w}}(x)} = \|H_{\bar{w}}(x)^{-1/2}v\|_{\bar{w}} \geq \frac{\|v\|_{\bar{w}}}{\|H_{\bar{w}}(x)\|_{\bar{w}}^{1/2}},$$

Theorem 3.5.11 applied with $\bar{w}$ in place of e thus gives

$$\|\eta s + g(x)\|_{-g(x)} \geq \frac{\|x + g_{\bar{w}}(x)\|_{\bar{w}}}{\|H_{\bar{w}}(x)\|_{\bar{w}}^{1/2}} \geq \min\left\{\tfrac{1}{5}, \tfrac{4}{5}\|x - \bar{w}\|_{\bar{w}}\right\},$$

completing the proof. □

The quantity $\|\eta s + g(x)\|_{-g(x)}$ is independent of inner products in the sense that if one expresses the original primal and dual instances in terms of a new inner product $\langle\,,\,\rangle_z$, transforming the dual feasible points by $s \mapsto H(z)^{-1}s$, one has

$$\|\eta s + g(x)\|_{-g(x)} = \|\eta H(z)^{-1}s + g_z(x)\|_{-g_z(x)},$$

as is easily verified using the identity $H(-g_z(x)) = H(z)H(-g(x))H(z)$. In the special case that $z = x$, this gives $\|\eta s + g(x)\|_{-g(x)} = \|\eta H(x)^{-1}s - x\|_x$, making evident that the quantity measures distance between the primal point and the (scaled) dual feasible point. Likewise, $\|\eta s + g(x)\|_{-g(x)} = \|\eta H(s)^{-1}x - s\|_s$.

3.7.2 An Algorithm

A feature of our first primal-dual path-following ipm is the simplicity of its analysis. Indeed, the algorithm is designed expressly for this feature. In the next subsection we present an algorithm which is more typical of the ipm's found in the literature, one that relies on the Nesterov–Todd directions. Its analysis extends the simpler analysis.

Assume x and s are the current feasible points, meant to crudely approximate x_η and s_η. Our first algorithm obtains better approximations by computing

$$x_+ := x + 2P_{L,\bar{w}}(\bar{w} - x) \quad \text{and} \quad s_+ := s + 2P_{L^\perp,\bar{w}^*}(\bar{w}^* - s),$$

where $P_{L,\bar{w}}$ denotes orthogonal projection onto L, orthogonal w.r.t. $\langle\,,\,\rangle_{\bar{w}}$. The algorithm then increases η to some value η_+. The points x_+, s_+—computed with the aim of obtaining better approximations to x_η, s_η than x, s—can be considered as crude approximations to x_{η_+}, s_{η_+}, just as x, s were considered to be crude approximations to x_η, s_η. The algorithm repeats the steps with η_+ in place of η, and so on.

Insight into the algorithm is gained by understanding that the primal and dual direction vectors

$$\Delta x := x_+ - x = 2P_{L,\bar{w}}(\bar{w} - x), \qquad \Delta s := s_+ - s = 2P_{L^\perp,\bar{w}^*}(\bar{w}^* - s)$$

can essentially be viewed as orthogonal projections of the same vector, similarly to the primal and dual Nesterov–Todd directions. In particular, we claim

$$H(w)^{-1}\Delta s = 2P_{L^{\perp_{\bar{w}}},\bar{w}}(\bar{w} - x), \tag{3.54}$$

and hence

$$\Delta x + H(w)^{-1}\Delta s = 2(\bar{w} - x)$$

gives an orthogonal decomposition of $2(\bar{w} - x)$.

To verify (3.54), first note that since $\bar{w}$ is a scalar multiple of w, the inner product $\langle\,,\,\rangle_{\bar{w}}$ is a scalar multiple of $\langle\,,\,\rangle_w$, and hence we have the operator identity $P_{L,\bar{w}} \equiv P_{L,w}$. Similarly, we have $P_{L^\perp,\bar{w}^*} \equiv P_{L^\perp,w^*}$. Relying on the operator identity $P_{L^\perp,w^*} \equiv H(w)P_{L^{\perp_w},w}H(w)^{-1}$ (i.e., relying on (3.48)), it follows that

$$P_{L^\perp,\bar{w}^*} \equiv H(w)P_{L^{\perp_{\bar{w}}},\bar{w}}H(w)^{-1}.$$

Also note that by logarithmic homogeneity (Theorem 2.3.9),

$$H(w)^{-1}\bar{w}^* = \tfrac{1}{\sqrt{\eta}}H(w)^{-1}w^*$$
$$= -\tfrac{1}{\sqrt{\eta}}H(w)^{-1}g(w)$$
$$= \tfrac{1}{\sqrt{\eta}}w$$
$$= \bar{w}.$$

Thus,

$$H(w)^{-1}\Delta s = 2H(w)^{-1}P_{L^\perp,\bar{w}^*}(\bar{w}^* - s)$$
$$= 2P_{L^{\perp_{\bar{w}}},\bar{w}}H(w)^{-1}(\bar{w}^* - s)$$
$$= 2P_{L^{\perp_{\bar{w}}},\bar{w}}(\bar{w} - x),$$

3.7. Primal-Dual Path-Following Methods

establishing (3.54) and the fact that

$$\Delta x + H(w)^{-1}\Delta s = 2(\bar{w} - x)$$

gives an orthogonal decomposition of $2(\bar{w} - x)$.

The orthogonal decomposition shows that Δx, Δs, and Δy form the (unique) solution of the system of linear equations

$$\begin{bmatrix} 0 & I & A_w^* \\ I & I & 0 \\ A & 0 & 0 \end{bmatrix} \begin{bmatrix} \Delta x \\ H(w)^{-1}\Delta s \\ \Delta y \end{bmatrix} = \begin{bmatrix} 0 \\ 2(\bar{w} - x) \\ 0 \end{bmatrix},$$

where $A_w^* = H(w)^{-1}A^* = \eta H(\bar{w})^{-1}A^* = \eta A_{\bar{w}}^*$. These equations can be used to compute the directions in practice.

In light of Theorem 3.7.1, the following theorem justifies the algorithm.

Theorem 3.7.2. *Assume* $|1 - \frac{\eta_+}{\eta}| \leq 1/(12\sqrt{\vartheta_f})$. *If x and s are feasible points satisfying*

$$\|\eta s + g(x)\|_{-g(x)} < \tfrac{1}{5},$$

then

$$\|\eta_+ s_+ + g(x_+)\|_{-g(x_+)} < \tfrac{1}{5}.$$

Proof. The final inequality in Theorem 3.7.1 implies

$$\|x - \bar{w}\|_{\bar{w}} < \tfrac{1}{4}.$$

Also note that since

$$\Delta x + H(w)^{-1}\Delta s = 2(\bar{w} - x),$$

we have

$$\bar{w} - x = \tfrac{1}{2}(\Delta x + H(w)^{-1}\Delta s),$$
$$\bar{w} - x_+ = \tfrac{1}{2}(-\Delta x + H(w)^{-1}\Delta s),$$

and

$$\bar{w} - H(w)^{-1}s_+ = \tfrac{1}{2}(\Delta x - H(w)^{-1}\Delta s).$$

Consequently,

$$H(w)^{-1}s_+ = 2\bar{w} - x_+,$$

and because $\Delta x \perp_{\bar{w}} H(w)^{-1}\Delta s$,

$$\|\bar{w} - x_+\|_{\bar{w}} = \|\bar{w} - x\|_{\bar{w}} < \tfrac{1}{4}.$$

Hence, by $g_{\bar{w}}(\bar{w}) = -\bar{w}$, $H_{\bar{w}}(\bar{w}) = I$, Corollary 1.5.8, and Theorem 2.2.1, we have

$$\|H(w)^{-1}s_+ + g_{\bar{w}}(x_+)\|_{\bar{w}} = \|g_{\bar{w}}(x_+) + 2\bar{w} - x_+\|_{\bar{w}}$$
$$= \|g_{\bar{w}}(x_+) - g_{\bar{w}}(\bar{w}) - H_{\bar{w}}(\bar{w})(x_+ - \bar{w})\|_{\bar{w}}$$
$$\leq \frac{\|x_+ - \bar{w}\|_{\bar{w}}^2}{1 - \|x_+ - \bar{w}\|_{\bar{w}}}$$
$$< \tfrac{1}{12}.$$

Since (3.39) (applied with $\bar{w}$ in place of e and x_+ in place of x) implies that for all v,

$$\|v\|_{-g_{\bar{w}}(x_+)} \leq (1 + \|x_+ - \bar{w}\|_{\bar{w}})\|v\|_{\bar{w}} \leq \tfrac{5}{4}\|v\|_{\bar{w}},$$

it follows that

$$\begin{aligned}
\|\eta s_+ + g(x_+)\|_{-g(x_+)} &= \|\eta H(\bar{w})^{-1} s_+ + g_{\bar{w}}(x_+)\|_{-g_{\bar{w}}(x_+)} \\
&= \|H(w)^{-1} s_+ + g_{\bar{w}}(x_+)\|_{-g_{\bar{w}}(x_+)} \\
&< \tfrac{5}{4} \cdot \tfrac{1}{12} \\
&= \tfrac{5}{48}.
\end{aligned}$$

Hence, using $|1 - \tfrac{\eta_+}{\eta}| \leq 1/(12\sqrt{\vartheta_f})$ and $\vartheta_f \geq 1$, we have

$$\begin{aligned}
\|\eta_+ s_+ + g(x_+)\|_{-g(x_+)} &\leq \tfrac{\eta_+}{\eta}\|\eta s_+ + g(x_+)\|_{-g(x_+)} + |1 - \tfrac{\eta_+}{\eta}|\,\|g(x_+)\|_{-g(x_+)} \\
&< \tfrac{13}{12} \cdot \tfrac{5}{48} + \tfrac{1}{12\sqrt{\vartheta_f}} \cdot \sqrt{\vartheta_f} \\
&< \tfrac{1}{5},
\end{aligned}$$

completing the proof. $\square$

Applying the basic step of the algorithm repeatedly, using $\eta_{i+1} := (1 + \tfrac{1}{12\sqrt{\vartheta_f}})\eta_i$, Theorem 3.7.2 guarantees the algorithm stays on track. Moreover, the value η will increase by a factor of at least 2 with every $14\sqrt{\vartheta_f}$ iterations, i.e., will increase by a factor of at least 2 within $O(\sqrt{\vartheta_f})$ iterations. Consequently, using Theorem 3.7.1, we see that the difference $\langle c, x_i \rangle - \langle b, y_i \rangle$ will decrease by a factor of at least $\tfrac{1}{2}$ within $O(\sqrt{\vartheta_f})$ iterations. This is essentially the same as what was proven for the barrier method in §2.4.2.

3.7.3 Another Algorithm

Here we present an algorithm which is typical of primal-dual path-following methods found in the literature, one that generalizes an approach for LP found in Kojima, Mizuno, and Yoshise [11] and in Monteiro and Adler [14]. It relies on the (scaled) Nesterov–Todd directions

$$\bar{d}_x := -P_{L,\bar{w}}(x + g_{\bar{w}}(x)) \qquad (= \tfrac{1}{\eta} d_x)$$

and

$$\bar{d}_s := -P_{L^\perp, \bar{w}^*}(s + g_{\bar{w}^*}(s)) \qquad (= \tfrac{1}{\eta} d_s)$$

introduced at the end of §3.6.

Given feasible x, s which crudely approximate x_η, s_η, the algorithm attempts to obtain better approximations to x_η, s_η by computing

$$x_+ := x + \bar{d}_x \qquad \text{and} \qquad s_+ := s + \bar{d}_s.$$

The algorithm then increases η to η_+. The points x_+, s_+ crudely approximate x_{η_+}, s_{η_+}. The algorithm repeats the steps with η_+ in place of η, and so on.

3.7. Primal-Dual Path-Following Methods

The algorithm is closely related to the one of the previous subsection. Indeed, if x, s are near x_η, s_η, then $\bar{w} \approx x$ and hence

$$x + g_{\bar{w}}(x) \approx x + g_{\bar{w}}(\bar{w}) + H_{\bar{w}}(\bar{w})(x - \bar{w}) = 2(x - \bar{w});$$

likewise, $s + g_{\bar{w}^*}(s) \approx 2(s - \bar{w}^*)$. Consequently, it is not surprising that the analysis of the algorithm is an extension of the arguments in the proof of Theorem 3.7.2.

Theorem 3.7.3. *Assume $|1 - \frac{\eta_+}{\eta}| \leq 1/(20\sqrt{\vartheta_f})$. If x and s are feasible points satisfying*

$$\|\eta s + g(x)\|_{-g(x)} < \tfrac{1}{10},$$

then

$$\|\eta_+ s_+ + g(x_+)\|_{-g(x_+)} < \tfrac{1}{10}.$$

Proof. The final inequality in Theorem 3.7.1 implies

$$\|x - \bar{w}\|_{\bar{w}} < \tfrac{5}{4} \cdot \tfrac{1}{10} = \tfrac{1}{8}.$$

Hence, defining

$$u := g_{\bar{w}}(x) + 2\bar{w} - x = g_{\bar{w}}(x) - g_{\bar{w}}(\bar{w}) - H_{\bar{w}}(\bar{w})(x - \bar{w}),$$

Corollary 1.5.8 and Theorem 2.2.1 give

$$\|u\|_{\bar{w}} \leq \frac{\|x - \bar{w}\|_{\bar{w}}^2}{1 - \|x - \bar{w}\|_{\bar{w}}} < \tfrac{1}{56}.$$

We know from §3.6 that

$$\bar{d}_x + H(w)^{-1} \bar{d}_s = -(x + g_{\bar{w}}(x))$$
$$= 2(\bar{w} - x) - u.$$

Hence,

$$\bar{w} - x = \tfrac{1}{2}(\bar{d}_x + H(w)^{-1}\bar{d}_s + u),$$
$$\bar{w} - x_+ = \tfrac{1}{2}(-\bar{d}_x + H(w)^{-1}\bar{d}_s + u),$$

and

$$\bar{w} - H(w)^{-1} s_+ = \tfrac{1}{2}(\bar{d}_x - H(w)^{-1}\bar{d}_s + u).$$

Consequently,

$$H(w)^{-1} s_+ = 2\bar{w} - x_+ - u. \tag{3.55}$$

Moreover, as we know from §3.6 that $\bar{d}_x \perp_{\bar{w}} H(w)^{-1}\bar{d}_s$,

$$\|\bar{w} - x_+\|_{\bar{w}} \leq \tfrac{1}{2}(\| - \bar{d}_x + H(w)^{-1}\bar{d}_s\|_{\bar{w}} + \|u\|_{\bar{w}})$$
$$= \tfrac{1}{2}(\|\bar{d}_x + H(w)^{-1}\bar{d}_s\|_{\bar{w}} + \|u\|_{\bar{w}})$$
$$\leq \|\bar{w} - x\|_{\bar{w}} + \|u\|_{\bar{w}}$$
$$< \tfrac{1}{8} + \tfrac{1}{56}$$
$$= \tfrac{1}{7}.$$

Hence, by (3.55), Corollary 1.5.8, and Theorem 2.2.1 we have

$$\begin{aligned}
\|H(w)^{-1}s_+ + g_{\bar{w}}(x_+)\|_{\bar{w}} &= \|g_{\bar{w}}(x_+) + 2\bar{w} - x_+ - u\|_{\bar{w}} \\
&\leq \|g_{\bar{w}}(x_+) + 2\bar{w} - x_+\|_{\bar{w}} + \|u\|_{\bar{w}} \\
&= \|g_{\bar{w}}(x_+) - g_{\bar{w}}(\bar{w}) - H_{\bar{w}}(\bar{w})(x_+ - \bar{w})\|_{\bar{w}} + \|u\|_{\bar{w}} \\
&\leq \frac{\|\bar{w} - x_+\|_{\bar{w}}^2}{1 - \|\bar{w} - x_+\|_{\bar{w}}} + \|u\|_{\bar{w}} \\
&< \tfrac{1}{42} + \tfrac{1}{56} \\
&= \tfrac{1}{24}.
\end{aligned}$$

However, (3.39) implies that for all v,

$$\|v\|_{-g_{\bar{w}}(x_+)} \leq (1 + \|\bar{w} - x_+\|_{\bar{w}})\|v\|_{\bar{w}} \leq \tfrac{8}{7}\|v\|_{\bar{w}}.$$

Thus,

$$\begin{aligned}
\|\eta s_+ + g(x_+)\|_{-g(x_+)} &= \|H(w)^{-1}s_+ + g_{\bar{w}}(x_+)\|_{-g_{\bar{w}}(x_+)} \\
&< \tfrac{8}{7} \cdot \tfrac{1}{24} \\
&= \tfrac{1}{21}.
\end{aligned}$$

Hence, using $|1 - \frac{\eta_+}{\eta}| \leq 1/(20\sqrt{\vartheta_f})$ and $\vartheta_f \geq 1$, we have

$$\begin{aligned}
\|\eta_+ s_+ + g(x_+)\|_{-g(x_+)} &\leq \tfrac{\eta_+}{\eta}\|\eta s_+ + g(x_+)\|_{-g(x_+)} + |1 - \tfrac{\eta_+}{\eta}|\|g(x_+)\|_{-g(x_+)} \\
&< \tfrac{21}{20} \cdot \tfrac{1}{21} + \tfrac{1}{20\sqrt{\vartheta_f}} \cdot \sqrt{\vartheta_f} \\
&= \tfrac{1}{10},
\end{aligned}$$

completing the proof. $\square$

3.8 A Primal-Dual Potential-Reduction Method

We assume K is a self-scaled cone and rely on the notation of §3.6. We continue to fix an arbitrary point $e \in K^\circ$ and let $\langle \, , \, \rangle = \langle \, , \, \rangle_e$.

3.8.1 The Potential Function

The progress of a potential reduction method is established by showing that a certain function—an appropriately chosen "potential function"—decreases by a constant amount with each iteration of the algorithm. The decrease can be established regardless of where the current iterate lies; it need not lie near the central path. Potential-reduction methods are not as restricted in their movement as path-following methods whose analysis depends on staying near the central path.

The primal-dual potential-reduction method we consider relies on the *Tanabe–Todd–Ye potential function* [20], [23]

$$\Phi(x, s) := \rho \ln\langle x, s\rangle + f(x) + f(s),$$

3.8. A Primal-Dual Potential-Reduction Method

where
$$\rho := \vartheta_f + \sqrt{\vartheta_f}.$$
The appropriateness of this potential function is shown by the following theorem, recalling that
$$\langle x, s \rangle = \langle c, x \rangle - \langle b, y \rangle.$$

Theorem 3.8.1. *For $x, s \in K^\circ$,*
$$\ln \langle x, s \rangle \leq \frac{1}{\sqrt{\vartheta_f}} (\Phi(x, s) - \vartheta_f \ln \vartheta_f - 2f(e))$$
(with equality iff s is a multiple of $-g(x)$).

Proof. Let $t := \vartheta_f / \langle x, s \rangle$. Recall that $f^*(z) = f(z) + C_e$ for all $z \in K^\circ$, where $C_e := -\vartheta_f - 2f(e)$ (Theorem 3.5.1). Thus, by definition of f^*,
$$\begin{aligned} f(ts) &= f^*(ts) + \vartheta_f + 2f(e) \\ &\geq -\langle x, ts \rangle - f(x) + \vartheta_f + 2f(e) \\ &= -f(x) + 2f(e) \end{aligned}$$
(with equality iff $ts = -g(x)$, i.e., iff s is a multiple of $-g(x)$). Consequently, since $f(ts) = f(s) - \vartheta_f \ln t$ (logarithmic homogeneity),
$$\begin{aligned} \Phi(x, s) &= (\vartheta_f + \sqrt{\vartheta_f}) \ln \langle x, s \rangle + f(x) + f(s) \\ &= \sqrt{\vartheta_f} \ln \langle x, s \rangle + \vartheta_f \ln \vartheta_f + f(x) + f(s) - \vartheta_f \ln t \\ &= \sqrt{\vartheta_f} \ln \langle x, s \rangle + \vartheta_f \ln \vartheta_f + f(x) + f(ts) \\ &\geq \sqrt{\vartheta_f} \ln \langle x, s \rangle + \vartheta_f \ln \vartheta_f + 2f(e) \end{aligned}$$
(with equality iff s is a multiple of $-g(x)$). $\square$

An immediate consequence of the theorem is that if an algorithm generates primal and dual feasible iterates for which $\Phi(x, s)$ goes to $-\infty$, then the objective values of the iterates converge to the optimal value.

The potential function is the sum of a strictly convex functional—the map $(x, s) \mapsto f(x) + f(s)$—and a functional which when restricted to the primal and dual feasible regions is concave. Indeed, to see that the functional
$$(x, s) \mapsto \rho \ln \langle x, s \rangle \tag{3.56}$$
is concave when restricted to the feasible regions, observe that its Hessian acts on vectors $(u, v) \in \mathbb{R}^n \times \mathbb{R}^n$ as follows:
$$\mathbf{H}(x, s) \begin{bmatrix} u \\ v \end{bmatrix} = \frac{\rho}{\langle x, s \rangle} \begin{bmatrix} v \\ u \end{bmatrix} - \frac{\rho \langle u, s \rangle + \rho \langle v, x \rangle}{\langle x, s \rangle^2} \begin{bmatrix} s \\ x \end{bmatrix}.$$
Thus, if $(u, v) \in L \times L^\perp$,
$$\left\langle \begin{bmatrix} u \\ v \end{bmatrix}, \mathbf{H}(x, s) \begin{bmatrix} u \\ v \end{bmatrix} \right\rangle = -\rho \left(\frac{\langle u, s \rangle + \langle v, x \rangle}{\langle x, s \rangle} \right)^2 \leq 0.$$
Hence, the Hessian of the functional obtained by restricting (3.56) to the primal and dual feasible regions is negative semidefinite, i.e., the restricted functional is concave.

3.8.2 The Algorithm

Assume $x \in K^\circ$ is primal feasible and $(y, s) \in \mathbb{R}^m \times K^\circ$ is dual feasible. The algorithm moves from these points to x_+ and (y_+, s_+). Maintaining our assumption that A is surjective—hence A^* is injective—y and y_+ are uniquely determined by s and s_+, respectively. Thus, to describe the algorithm, we need only describe how x_+ and s_+ are computed from x and s. We hide y and y_+ to make the symmetry of the algorithm apparent.

The points x_+ and s_+ arise from a line search using the Nesterov–Todd directions

$$d_x := -P_{L,w}(\eta x + g_w(x)) \quad \text{and} \quad d_s := -P_{L^\perp, w^*}(\eta s + g_{w^*}(s)),$$

where

$$\eta := \frac{\rho}{\langle x, s \rangle}.$$

For motivation, we note that the gradient w.r.t. $\langle \ , \ \rangle_w$ of the functional obtained by restricting $x \mapsto \Phi(x, s)$ to the primal feasible region is precisely $-d_x$, i.e., d_x is the direction of steepest descent. Likewise, d_s is the direction of steepest descent w.r.t. $\langle \ , \ \rangle_{w^*}$ for the functional obtained by restricting $s \mapsto \Phi(x, s)$ to the dual feasible region.

The algorithm searches the line $\{(x + td_x, s + td_s) : t \in \mathbb{R}\}$ to obtain an approximate minimizer $t_{\min}$ of the univariate functional

$$\phi(t) := \Phi(x + td_x, s + td_s).$$

It then lets

$$(x_+, s_+) := (x + t_{\min} d_x, s + t_{\min} d_s).$$

We show

$$\inf_t \phi(t) < \phi(0) - \tfrac{1}{250}.$$

For theory purposes we can formalize the condition that $t_{\min}$ be an "approximate minimizer" by requiring $t_{\min}$ satisfy, say,

$$\phi(t_{\min}) \leq \phi(0) - \tfrac{1}{500}. \tag{3.57}$$

The specific constant $\frac{1}{500}$ is irrelevant for the theory. What matters is that the constant is positive and universal (i.e., entirely independent of the particular instances) and that potential reduction of the amount can always be achieved. In practice, drastically greater reduction of the potential function than $\frac{1}{500}$ always happens.

The algorithm generates a sequence of points $(x_0, s_0), (x_1, s_1), \ldots$, each obtained from the previous one by line search. The relation (3.57) in conjunction with Theorem 3.8.1 immediately imply for all i that

$$\ln \langle x_i, s_i \rangle \leq \frac{1}{\sqrt{\vartheta_f}} \left(\Phi(x_0, s_0) - \vartheta_f \ln \vartheta_f - 2f(e) - \tfrac{1}{500} i \right).$$

Hence, for all $\epsilon > 0$, the algorithm computes feasible points (x_i, s_i) satisfying

$$\langle x_i, s_i \rangle \leq \epsilon$$

within

$$500(\sqrt{\vartheta_f} \ln \tfrac{1}{\epsilon} + \Phi(x_0, s_0) - \vartheta_f \ln \vartheta_f - 2f(e)) \tag{3.58}$$

3.8. A Primal-Dual Potential-Reduction Method

iterations.

From a theoretical perspective, by far the most important term in the bound (3.58) is $\sqrt{\vartheta_f} \ln \frac{1}{\epsilon}$. The bound can be roughly interpreted as stating that the objective value difference $\langle c, x_i \rangle - \langle b, y_i \rangle$ is halved in $O(\sqrt{\vartheta_f})$ iterations, a result which is essentially the same as what we obtained for the barrier method in §2.4.2 and for the primal-dual path-following methods in §3.7. However, the potential-reduction method has the potential for much greater efficiency (especially if each line search yields $t_{\min}$ as the exact minimizer) as it is not required to stay near the central path like the other algorithms.

The potential-reduction method described above was first introduced in the context of linear complementarity problems by Kojima, Mizuno, and Yoshise [12].

3.8.3 The Analysis

It remains to prove the following theorem.

Theorem 3.8.2.
$$\inf_t \phi(t) \leq \phi(0) - \tfrac{1}{250}.$$

Proof. Recall from §3.6 the definitions
$$\bar{w} := \tfrac{1}{\sqrt{\eta}} w, \qquad \bar{w}^* := \tfrac{1}{\sqrt{\eta}} w^*,$$

$$\bar{d}_x := -P_{L,\bar{w}}(x + g_{\bar{w}}(x)) = \tfrac{1}{\eta} d_x \quad \text{and} \quad \bar{d}_s := -P_{L^\perp, \bar{w}^*}(s + g_{\bar{w}^*}(s)) = \tfrac{1}{\eta} d_s.$$

Also recall that
$$\bar{d}_x + H(w)^{-1} \bar{d}_s = -(x + g_{\bar{w}}(x)) \tag{3.59}$$

is an orthogonal (w.r.t. $\langle \ , \ \rangle_{\bar{w}}$) decomposition of $-(x + g_{\bar{w}}(x))$.

Observe that
$$\|x\|_{\bar{w}}^2 = \langle x, H(\bar{w}) x \rangle = \eta \langle x, s \rangle.$$

Hence, since $\eta := \rho / \langle x, s \rangle$,
$$\|x\|_{\bar{w}} = \sqrt{\rho} \quad \text{and, likewise,} \quad \|s\|_{\bar{w}^*} = \sqrt{\rho}. \tag{3.60}$$

For convenience, we scale the search directions, defining
$$(\check{d}_x, \check{d}_s) := \frac{1}{\|H_{\bar{w}}(x)\|_{\bar{w}}^{1/2} \|x + g_{\bar{w}}(x)\|_{\bar{w}}} (\bar{d}_x, \bar{d}_s). \tag{3.61}$$

Since (3.59) is an orthogonal decomposition,
$$\|\check{d}_x\|_{\bar{w}}, \|H(w)^{-1} \check{d}_s\|_{\bar{w}} \leq 1 / \|H_{\bar{w}}(x)\|_{\bar{w}}^{1/2}. \tag{3.62}$$

Let
$$\check{\phi}(t) := \Phi(x + t \check{d}_x, s + t \check{d}_s).$$

It suffices to prove
$$\check{\phi}(\tfrac{1}{25}) \leq \check{\phi}(0) - \tfrac{1}{250}. \tag{3.63}$$

For all $\tilde{x}, \tilde{s} \in K^\circ$, define
$$\Phi_{\tilde{w}}(\tilde{x}, \tilde{s}) := \rho \ln \langle \tilde{x}, \tilde{s} \rangle_{\tilde{w}} + f(\tilde{x}) + f(\tilde{s}).$$

Let
$$\check{\phi}_{\tilde{w}}(t) := \Phi_{\tilde{w}}(x + t\check{d}_x, x + tH(w)^{-1}\check{d}_s).$$

Since for all $\tilde{x}, \tilde{s} \in K^\circ$,
$$\ln \langle \tilde{x}, \tilde{s} \rangle = \ln \langle \tilde{x}, H(w)^{-1}\tilde{s} \rangle_{\tilde{w}} - \ln \eta$$

and, using (3.34),
$$f(\tilde{s}) = f(H(w)^{-1}\tilde{s}) + 2(f(e) - f(w^*)),$$

we have
$$\Phi(\tilde{x}, \tilde{s}) = \Phi_{\tilde{w}}(\tilde{x}, H(w)^{-1}\tilde{s}) - \rho \ln \eta + 2(f(e) - f(w^*)),$$

and hence,
$$\check{\phi}(t) = \check{\phi}_{\tilde{w}}(t) - \rho \ln \eta + 2(f(e) - f(w^*)).$$

Consequently, to establish (3.63) it suffices to show
$$\check{\phi}_{\tilde{w}}(\tfrac{1}{25}) \le \check{\phi}_{\tilde{w}}(0) - \tfrac{1}{250}.$$

Let
$$\psi_1(t) := \rho \ln \langle x + t\check{d}_x, x + tH(w)^{-1}\check{d}_s \rangle_{\tilde{w}} \quad (= \rho \ln \langle x + t\check{d}_x, s + t\check{d}_s \rangle + \rho \ln \eta),$$
$$\psi_2(t) := f(x + t\check{d}_x), \quad \text{and} \quad \psi_3(t) := f(x + tH(w)^{-1}\check{d}_s),$$

so that
$$\check{\phi}_{\tilde{w}}(t) = \psi_1(t) + \psi_2(t) + \psi_3(t).$$

The function ψ_1 is concave since the map $(x, s) \to \rho \ln \langle x, s \rangle$ is concave when restricted to the primal and dual feasible regions (as was discussed at the end of §3.8.1). Hence,
$$\psi_1(t) \le \psi_1(0) + t\psi_1'(0),$$

that is,
$$\psi_1(t) \le \psi_1(0) + t \frac{\rho}{\|x\|_{\tilde{w}}^2} \langle x, \check{d}_x + H(w)^{-1}\check{d}_s \rangle_{\tilde{w}}.$$

Thus, using (3.59), (3.60), and (3.61), we have
$$\psi_1(t) \le \psi_1(0) - t \cdot \frac{\langle x, x + g_{\tilde{w}}(x) \rangle_{\tilde{w}}}{\|H_{\tilde{w}}(x)\|_{\tilde{w}}^{1/2} \|x + g_{\tilde{w}}(x)\|_{\tilde{w}}}.$$

Using (3.62) we have
$$\|\check{d}_x\|_x \le \|H_{\tilde{w}}(x)\|_{\tilde{w}}^{1/2} \|\check{d}_x\|_{\tilde{w}} \le 1.$$

3.8. A Primal-Dual Potential-Reduction Method

Consequently, by Theorem 2.2.2, for $0 \le t < 1$,

$$\psi_2(t) \le \psi_2(0) + t\langle g_x(x), \check{d}_x\rangle_x + \tfrac{t^2}{2} + \tfrac{t^3}{3(1-t)}$$
$$\le \psi_2(0) + t\langle g_x(x), \check{d}_x\rangle_x + \tfrac{t^2}{1-t}$$
$$= \psi_2(0) + t\langle g_{\bar{w}}(x), \check{d}_x\rangle_{\bar{w}} + \tfrac{t^2}{1-t}.$$

Similarly, $\|H(w)^{-1}\check{d}_s\|_x \le 1$ and hence

$$\psi_3(t) \le \psi_3(0) + t\langle g_{\bar{w}}(x), H(w)^{-1}\check{d}_s\rangle_{\bar{w}} + \tfrac{t^2}{1-t}.$$

Thus, using (3.59) and (3.61),

$$\psi_2(t) + \psi_3(t) \le \psi_2(0) + \psi_3(0) - t \cdot \frac{\langle g_{\bar{w}}(x), x + g_{\bar{w}}(x)\rangle_{\bar{w}}}{\|H_{\bar{w}}(x)\|_{\bar{w}}^{1/2}\|x + g_{\bar{w}}(x)\|_{\bar{w}}} + \frac{2t^2}{1-t}.$$

In all, for $0 \le t < 1$,

$$\check{\phi}_{\bar{w}}(t) = \psi_1(t) + \psi_2(t) + \psi_3(t)$$
$$\le \check{\phi}_{\bar{w}}(0) - t \cdot \frac{\|x + g_{\bar{w}}(x)\|_{\bar{w}}}{\|H_{\bar{w}}(x)\|_{\bar{w}}^{1/2}} + \frac{2t^2}{1-t}.$$

Hence, by Theorem 3.5.11,

$$\check{\phi}_{\bar{w}}(t) \le \check{\phi}_{\bar{w}}(0) - t \cdot \min\left\{\tfrac{1}{5}, \tfrac{4}{5}\|x - \bar{w}\|_{\bar{w}}\right\} + \tfrac{2t^2}{1-t}.$$

However, since $\|x\|_{\bar{w}}^2 = \rho = \vartheta_f + \sqrt{\vartheta_f}$, $\|\bar{w}\|_{\bar{w}} = \sqrt{\vartheta_f}$, and $\vartheta_f \ge 1$, it is not difficult to prove that

$$\|x - \bar{w}\|_{\bar{w}} \ge \tfrac{1}{4}.$$

Thus,

$$\check{\phi}_{\bar{w}}(t) \le \check{\phi}_{\bar{w}}(0) - \tfrac{t}{5} + \tfrac{2t^2}{1-t}.$$

Substitution yields

$$\check{\phi}_{\bar{w}}(\tfrac{1}{25}) \le \check{\phi}_{\bar{w}}(0) - \tfrac{1}{250},$$

concluding the proof. □

Bibliography

[1] E.J. Andersen and P. Nash, *Linear Programming in Infinite Dimensional Spaces: Theory and Applications*, Wiley, Chichester, 1987.

[2] H.H. Bauschke, O. Güler, A.S. Lewis, and H.S. Sendov, "Hyperbolic polynomials and convex analysis," *Canadian Journal of Mathematics*, 53 (2001), pp. 470–488.

[3] D. den Hertog, C. Roos, and J.-Ph. Vial, "A complexity reduction for the long-step path-following algorithm for linear programming," *SIAM Journal on Optimization*, 2 (1992), pp. 71–87.

[4] J. Faraut and A. Koranyi, *Analysis on Symmetric Cones,* Clarendon Press, Oxford, 1994.

[5] L. Faybusovich, "Euclidean Jordan algebras and interior-point algorithms," *Positivity*, 1 (1997), pp. 331–335.

[6] C. Gonzaga, "An algorithm for solving linear programming problems in $O(n^3 L)$ operations," in *Progress in Mathematical Programming—Interior Point and Related Methods,* N. Megiddo, ed., Springer-Verlag, Berlin, 1989, pp. 1–28.

[7] C.C. Gonzaga, "Large step path-following methods for linear programming I. Barrier function method," *SIAM Journal on Optimization*, 1 (1991), pp. 268–279.

[8] O. Güler, "Barrier functions in interior point methods," *Mathematics of Operations Research*, 21 (1996), pp. 860–885.

[9] O. Güler, "Hyperbolic polynomials and interior point methods for convex programming," *Mathematics of Operations Research*, 22 (1997), pp. 350–377.

[10] N. Karmarkar, "A new polynomial time algorithm for linear programming," *Combinatorica*, 4 (1984), pp. 373–395.

[11] M. Kojima, S. Mizuno, and A. Yoshise, "A primal-dual interior point algorithm for linear programming," in *Progress in Mathematical Programming—Interior Point and Related Methods*, N. Megiddo, ed., Springer-Verlag, Berlin, 1989, pp. 29–47.

[12] M. Kojima, S. Mizuno, and A. Yoshise, "An $O(\sqrt{n}L)$ potential reduction method for linear complementarity problems," *Mathematical Programming*, 50 (1991), pp. 331–342.

[13] N. Megiddo, "Pathways to the optimal set in linear programming," in *Progress in Mathematical Programming—Interior Point and Related Methods,* N. Megiddo, ed., Springer-Verlag, Berlin, 1989, pp. 131–158.

[14] R.D.C. Monteiro and I. Adler, "Interior path following primal-dual algorithms. Part I: Linear programming," *Mathematical Programming,* 44 (1989), pp. 27–41.

[15] Yu.E. Nesterov and A.S. Nemirovskii, *Interior-Point Polynomial Algorithms in Convex Programming,* SIAM, Philadelphia, 1994.

[16] Yu.E. Nesterov and M.J. Todd, "Self-scaled barriers and interior-point methods for convex programming," *Mathematics of Operations Research,* 22 (1997), pp. 1–42.

[17] Yu.E. Nesterov and M.J. Todd, "Primal-dual interior-point methods for self-scaled cones," *SIAM Journal on Optimization,* 8 (1998), pp. 324–364.

[18] J. Renegar, "A polynomial-time algorithm based on Newton's method for linear programming," *Mathematical Programming,* 40 (1988), pp. 59–94.

[19] R.T. Rockafellar, *Convex Analysis,* Princeton University Press, Princeton, NJ, 1970.

[20] K. Tanabe, "Centered Newton method for mathematical programming," *System Modelling and Optimization: Proceedings of the* 13*th IFIP Conference,* Tokyo, Japan, Aug./Sept. 1987, *Lecture Notes in Control and Information Sciences* 113, M. Iri and K. Yajima, eds., Springer-Verlag, Berlin, 1988, pp. 197–206.

[21] T. Terlaky (editor), *Interior Point Methods of Mathematical Programming,* Kluwer, Dordrecht, The Netherlands, 1996.

[22] M.J. Todd, "Potential-reduction methods in mathematical programming," *Mathematical Programming,* 76 (1996), pp. 3–45.

[23] M.J. Todd and Y. Ye, "A centered projective algorithm for linear programming," *Mathematics of Operations Research,* 15 (1990), pp. 508–529.

[24] S.J. Wright, *Primal-Dual Interior-Point Methods,* SIAM, Philadelphia, 1997.

[25] Y. Ye, *Interior Point Algorithms: Theory and Analysis,* John Wiley and Sons, New York, 1997.

Index

A^*, 3
A_x^*, 22
analytic center, 39
asymptotic feasibility, 71
asymptotic optimal value, 71, 73
a-val, 71
a-val*, 73

$B_x(y, r)$, 23
barrier functional, 35

central path, 43, 81
complexity value, 35
ϑ_f, 35
conjugate functional, 76
f^*, 76
$\mathcal{C}^1$, 6
$\mathcal{C}^2$, 9

$x_1 \cdot x_2$, 2
dual cone, 65
K^*, 65
K_z^*, 83

Frobenius norm, 2

gradient, 6
$g(x)$, 6
$g_x(y)$, 22
$g|_{L,x}(y)$, 22

Hessian, 9
$H(x)$, 9
$H|_L(x)$, 11
$H_x(y)$, 22
$H|_{L,x}(y)$, 22

$\langle\ ,\ \rangle$, 2
$\langle\ ,\ \rangle_s$, 5

$\langle\ ,\ \rangle_x$, 22
intrinsically self-conjugate, 83
intrinsically self-dual, 83

logarithmic homogeneity, 42

Nesterov–Todd directions, 98, 99
$n(x)$, 19
$\|\ \|$, 2
$\|\ \|_s$, 5
$\|\ \|_x$, 22
$\|A\|$, 4

positive definite (pd), 4
positive semidefinite (psd), 4
P_L, 8
$P_{L,x}$, 22

$q_x(y)$, 18

$\mathbb{R}^n_{++}$, 24

scaling point, 86
self-adjoint, 4
self-concordant functional, 23
self-scaled (or symmetric) cone, 84
$\mathcal{SC}$, 23
$\mathcal{SCB}$, 35
strong duality, 67
strong feasibility, 73
$\mathbb{S}^{n \times n}$, 2
$\mathbb{S}^{n \times n}_{++}$, 6
sym(x, D), 40

$X_1 \circ X_2$, 2

val, 65
val*, 66

weak duality, 66